# Walt's Wisdom

## A Kentucky Cornucopia of Gardening Miscellany

*11-15-14*
*To Mike,*
*Best wishes and happy gardening.*
*Walt*

# Walt's Wisdom

## A Kentucky Cornucopia of Gardening Miscellany

**Walt Reichert**

Kentucky Monthly

*Walt's Wisdom: A Kentucky Cornucopia of Gardening Miscellany* is published by
Vested Interest Publications and *Kentucky Monthly*.

**Written by** Walt Reichert

**Publisher** Stephen M. Vest
**Executive Editor** Kim Butterweck
**Associate Editor** Patricia Ranft
**Contributing Editor** Madelynn Coldiron

**Art Director and Designer** Kelli Schreiber
**Contributing Photographer** Beth McCoy

**Director of Retail Sales** Kendall Shelton

Copyright© 2014 by Vested Interest Publications/*Kentucky Monthly*
All rights reserved. No part of this book may be copied or reproduced without permission from the publisher, except by a reviewer who may quote brief passages in a review.
Manufactured in partnership with Four Colour Print Group of Louisville, Ky.; printed in Korea.

ISBN 13: 978-0-615-98432-2
Price: $19.95

All book order correspondence should be addressed to:

**Kentucky Monthly**

Kentucky Monthly
P.O. Box 559
Frankfort, KY 40602-0559

1-888-329-0053
www.kentuckymonthly.com

This book is dedicated to my parents, Carolyn and Stu Reichert, who nurtured my gardening and writing interests, and to my wife, Mary Lou, and children Michael, Elizabeth and Rebecca.

# Introduction

This book is a compilation of gardening advice, ruminations, informed (I hope!) suggestions and profiles published in *Kentucky Monthly*'s gardening column over the last 13 years.

While 13 years is a long tenure in the publishing business, my connection with *Kentucky Monthly* actually goes back to the magazine's original incarnation in the mid-1970s. I wrote a feature story for the publication's second issue and shortly after started writing a garden column. That column—and the magazine—lasted a few years. Copies of those original *Kentucky Monthly* magazines have been found in antique stores. Talk about making a guy feel old!

But I can't escape the fact that I'm in my fifth decade of gardening and my fourth decade of writing about it. The earliest gardening experience I can remember is following my uncle "Red" out to the tomato patch with my sister and a hoe in tow. That gardening experience got cut short, however, when Red invited us to go back into the house; seems we were cutting down tomato plants and hilling up weeds! Other early gardening recollections were following my grandmother through the okra patch and pulling the sticky fruits from plants growing well above my head; riding on my grandfather's Allis-Chalmers tractor when he plowed for the spring garden; waiting for my great-grandmother's bed of pink and white peonies to bloom, which marked the end of the school year; and watching my dad spend summer evenings fussing over his hybrid tea roses.

Looking back, it doesn't seem long after those early recollections that I started gardening myself. My favorite thing to grow was watermelons—still is! And then suddenly I was out of college and writing about gardening, something I enjoy doing almost as much as playing in the dirt. My first jobs were teaching English in high school and college, but when I wasn't in class or grading papers, I was either gardening or writing about it. I had columns in *Kentucky Monthly* and then *Dayton* magazines before I was asked by then-editor Michael Embry to write for the current incarnation of *Kentucky Monthly*. As reporter and editor of the *Sentinel-News* in

Shelbyville for 10 years, I took full advantage of every opportunity to write about and take pictures of local gardens and gardeners. Many of the photographs that appeared in *Kentucky Monthly* columns came from the beautiful grounds of Shelby County gardeners who were gracious enough to invite me to "wander around whenever you want to."

I'm now on my third career (and still writing *Kentucky Monthly*'s gardening column) as the horticulture technician for the Shelby County Cooperative Extension Service. I spend my days visiting local gardens and gardeners, diagnosing plant problems, teaching gardening classes and working once a week in a public garden, the Tim McClure Botanical Gardens in Shelbyville. I'm surrounded by plants, gardens and good people—and they even pay me!

What I've learned on this job is that, as I expected, there is enormous interest in gardening. In Shelby County, at least, and I expect across the state, there is strong and increasing interest in growing food and fruit in home gardens. Yet the garden wisdom our parents and grandparents took for granted as part of growing up isn't there for this generation. But the hunger for knowledge is there. When I offered to teach a class on food gardening in the summer of 2013, the response was so overwhelming I had to open up two classes, and even then we had to turn some people away because of space limitations. I've found that folks are always searching for down-to-earth gardening information and advice they can follow in their food and ornamental gardens and landscapes. Unfortunately, so many books, magazines and websites that address gardening are national in scope, and the garden wisdom doesn't always apply to Kentucky gardens.

In the pages of this book, I hope gardeners can find at least some of the answers they're looking for. The columns are arranged in seasonal sequence, starting, naturally, with spring. Spring is the time of hope for gardeners—that catalog-picture-perfect garden can still happen. The attention then turns to summer, when activity is most feverish and we sometimes realize that the "best laid plans …" Then we move on to fall, when the respite from heat and humidity allows for moments of actual enjoyment in the garden and time to reflect on what did and didn't work out according to plan. We end with winter, when the gardener can hopefully recharge the brain and body and start getting the itch to do it all over again. The sidebars and tips, written most recently, are offered to update some of the stories and to add digestible bits of advice that flavor the original column.

It is my hope that this collection ignites your love of gardening and keeps the fire burning throughout the four seasons of your Kentucky garden.

# Acknowledgments

    I want to thank the staff of *Kentucky Monthly* magazine, especially Steve Vest and Kim Butterweck, for their untiring efforts in seeing this book through to completion. I appreciate their creativity, skill and good humor. Also to the magazine's editors and publishers, past and present, for allowing me to write about gardens and gardening in the pages of *Kentucky Monthly* for so many years.

    I also would like to thank my family for putting up with my gardening (and other) obsessions. Finally, I'd like to thank the staff of the Shelby County Cooperative Extension Service for allowing me to have a job that lets me play among plants and gardens all day.

# Contents

## SPRING

| | |
|---|---|
| Plant those cole crops | 13 |
| Promise yourself a rose garden | 16 |
| Grow the signature shrubs of the South | 20 |
| Royal purple in the garden | 23 |
| Back to the garden | 27 |
| Tips to make your garden greener | 30 |
| Do justice to your roses—Monty's way | 34 |
| Try something new | 37 |
| In love with lilacs | 40 |
| This Judas won't betray you | 43 |
| Too busy to garden? Try carefree plants | 46 |
| Honeymoon time in the garden | 49 |
| Five fabulous perennials | 51 |
| Hail, Iris, goddess of the rainbow | 53 |
| Suit yourself | 56 |
| 10 tips for your best-ever vegetable garden | 58 |
| Made in the shade | 61 |
| First fruits: strawberries | 64 |
| Marry plants for a razzle-dazzle garden | 67 |
| Homegrown corn: sweet! | 70 |

| | |
|---|---:|
| Enjoy the fruits of your labor | 74 |
| A flair for ferns | 78 |
| Create your own garden island | 81 |

## SUMMER

| | |
|---|---:|
| Pass the veggies, hold the poisons | 85 |
| Let the sunflowers shine | 88 |
| Gardening inside and out | 91 |
| Don't slime okra | 94 |
| Try squash without the guilt | 97 |
| Q&A: What's bugging you this summer? | 100 |
| Rein in your garden bullies | 103 |
| Sage gardeners plant salvia | 106 |
| Take a walk in the Wiche garden | 109 |
| Soar with the queen of climbers | 112 |
| Take it easy out there | 114 |
| Make room for flying flowers | 116 |
| Water, water everywhere | 119 |
| Growing tomatoes | 121 |
| Vines: gardening up, not out | 124 |
| 10 totally terrific tomatoes | 127 |
| Invite insects | 130 |
| Plant the lilies of summer | 132 |
| Take the melon challenge | 135 |

## FALL

| | |
|---|---:|
| Berry nice | 139 |
| Time to paint the spring garden | 142 |
| Finish that fall checklist | 146 |
| It's September, think spring | 149 |
| How to pass the soil test | 151 |
| Live and let live | 153 |
| Any way you say it, peonies offer lasting beauty | 155 |
| Magnolia, magnificent! | 157 |

| | |
|---|---|
| Reflections on a dismal season | 159 |
| Make room for big trees | 161 |
| Get the blues out of bluegrass | 164 |
| Hydrangea, a plant for every season | 166 |
| Bernheim Arboretum and Research Forest | 170 |
| Plant a green screen | 172 |
| Foreign invaders attack our state's forests | 175 |

# WINTER

| | |
|---|---|
| New Yorkers high on Kentucky gardener Jon Carloftis | 179 |
| Bark up the right tree | 181 |
| Is your garden for the birds? | 184 |
| Catch up on your reading | 187 |
| Match the garden to your cottage | 189 |
| The school of hard knocks | 192 |
| Garden catalogs' temptations | 196 |
| Caution: orchids can be addictive | 198 |
| Tool time's no joke | 200 |
| Gardens aren't just for plants | 202 |
| A plant by any other name ... | 204 |
| Gardeners will save the world | 206 |
| Five garden questions | 208 |
| HELP! Squelch invasive plants | 211 |
| Are you mooning? | 214 |

# Spring

# Plant those cole (not cold) crops

As an English teacher and former newspaper editor, I confess that certain misspellings drive me crazy. For example, I can't stand to see "a lot" spelled as one word, "alot," even though, since 90 percent of my students make that mistake, you'd think I'd be used to it.

Every spring one of the local garden centers slaps a misspelling on its sign that also gets on my last nerve. The sign advertises "cold" crop plants—in other words, broccoli, cabbage and cauliflower. But it's not "cold" crops; it's "cole" crops. "Cole" as in coleslaw. Get it?

Misspelled or not, cole crops belong in every vegetable garden. The cole crops include not only broccoli, cabbage and cauliflower, but also Brussels sprouts, collards, Chinese cabbage, kale and kohlrabi. The first three are by far the most commonly grown and eaten, at least in this part of the country. But kale can be harvested almost all winter, and collards tolerate our hot summers well, so be willing to give them a try.

With a couple of exceptions, cole crops are not difficult to grow, and they are far more flavorful fresh from the garden than from your grocer's shelf—by way of California, Florida or Chile. The cole crops are chock-full of vitamins and minerals that, scientists say, protect everything from your heart to your head. The most nutritious of the bunch, broccoli, for example, has large amounts of vitamins A, C and some Bs and also plenty of iron and calcium. Broccoli and cauliflower freeze well and cabbage can keep for months in a cool, humid place, like the bottom drawer of the refrigerator. Cabbage can also be frozen for use in winter soups.

## Planting and Growing

Probably the reason cole crops get misspelled as cold crops is they do grow best when it's cool, though they don't like it really cold. That's why they are often grown in winter in warm climates. Temperatures between 45 and 65 suit them best. In most of Kentucky, broccoli, cabbage and cauliflower ought to be planted no later than mid-April. You want to time planting to avoid freezes, but also have the plants grown before the sweltering heat of summer sets in. Cole crops will tolerate a light frost, but a heavy frost sets them back, and a freeze will kill them. If you plant in March and a freak cold spell arrives, just cover the plants with old bushel baskets or boxes and they'll usually weather the storm. Of all the cole crops, cauliflower is the most temperamental and will perform poorly if the weather turns either too cold or too hot. I always know it's been a good year for cole crops when I get a harvest of decent-size heads of cauliflower. Happens about once every three years.

The trick to growing cole crops is to keep them growing quickly. Set the plants about 18 inches apart—more for the larger cultivars of cabbage—in the richest soil you have. If your soil is poor, dig in some bagged cow manure or fertilize with 2 to 4 pounds of granulated fertilizer, such as 8-12-8, about every 25 feet of row. Since cole crops are big drinkers, you might consider mulching them to retain moisture. Usually our springs are wet enough, though; keeping plants from drowning is a more typical problem. If your garden drains poorly, think about planting cole crops in raised beds, where they perform extraordinarily well.

## Bugs

One of the reasons for getting cole crops in early, while the weather is cool, is that their most common insect pests have not awakened yet. Probably the most devastating attack comes from the cabbage maggot, which, despite its name, is more likely to attack broccoli and cauliflower than cabbage. The cabbage maggot is the larvae of a fly that lays eggs at the base of the plant. When the eggs hatch, the maggots eat the roots. The result is a beautiful patch of broccoli one day, and a wilted, dying mess the next day. The problem can be solved by using floating row covers that screen out the flies, but I've found the attack so rare that it's hardly worth the trouble.

A more common insect pest of broccoli, cabbage and cauliflower is the so-called cabbageworm, which can be either the imported cabbageworm or the cabbage looper. The former is, by far, the more common. The cabbageworm is a serious nuisance because it eats holes in the leaves, thereby reducing yield, and, most disgustingly, hides itself among the "branches" of broccoli, where it is almost invisible. Nothing like biting into a fresh broccoli floret and getting way more protein than you bargained for in the form of a little green worm.

If you see little white butterflies hovering over your cole crops, it's time to take action against the cabbageworm. You can cover the plants with lightweight row covers, as mentioned. Or you can use an insecticide containing an organic ingredient, Bt for short, that will kill the worms but does not harm other insects. Because Bt is short-lived, it needs to be applied frequently. Insecticides containing rotenone, pyrethrum, carbaryl and malathion also will kill cabbageworms, but be sure to spray the undersides of the leaves, where the worms are most common. Follow label directions for application.

## Harvest

Cabbage, cauliflower, Brussels sprouts and kohlrabi are cut-one-time crops. White cauliflower will turn purple if exposed to too much sun; to avoid that, tear off an outer leaf when the developing head is about the size of a silver dollar and cover it. Just be careful to keep an eye on development; the heads grow quickly and soon deteriorate if not picked right at their prime. Pick cabbage when the head is firm.

If the soil is rich, broccoli will keep producing small heads for weeks after the main head is cut. But be vigilant about keeping the heads cut out; if not harvested in a timely manner, broccoli will bolt (go to seed), and then the plant's done.

Many people also plant cole crops in late summer for a fall crop. But I've found the timing is tricky. Plant too soon in summer and the plants poop out in the heat; plant too late and early frosts take them out. Insects are also more of a problem in late summer plantings. Kale is probably the best of the coles for making a fall crop.

## FLUSH CUTWORMS

Many gardeners complain that they lose cole crop plants—and other plants, including tomatoes and peppers—to something that appears to have just chopped them down. That's the work of the cutworm. Cutworms are especially troublesome in gardens that have been in sod for years before being turned into vegetable gardens. Cutworms crawl out of the ground at night, chew on the plant stems and timberrr! To eliminate cutworm damage, you can dust the plants with insecticide immediately after planting. Or make use of those little cardboard rolls toilet paper is wrapped around. Put the plant through the roll as you are putting the plant in the ground. The roll will protect the plant's stem from damage and degrade into the soil as the season progresses.

# Promise yourself a rose garden

Why is it that problem children are often loved the best? Take hybrid tea roses. Hybrid tea roses outsell all the other dozen or so species of roses combined even though they may require more coddling than all of the others put together. Old roses are hardier; damask and gallica roses smell sweeter; shrub roses blend into the landscape better; miniature roses are cuter; grandifloras produce way more flowers.

And yet. And yet the rose we think of when we think of "rose" as in "blanket of roses," "a rose by any other name" or "I never promised you a rose garden" is the hybrid tea.

Perhaps because of their large, single, upright blooms that have been crafted over the last 100-plus years into a rainbow of colors and color blends, hybrid teas are the roses we plant by the hundreds of thousands every year, even though they may be the most challenging to grow.

My dad has raised roses most of his life. While I will try any rose a company will send me to trial, my dad will let only the hybrid tea grace his garden. Perhaps that is because after 50-plus years of growing roses and following in the footsteps of his dad before him, my dad has growing roses—hybrid teas—down to a science. It is not uncommon for him to have a rose with 30 blooms to a bush.

I can't promise you a rose garden that looks like my dad's if you try his tricks. But if you have given up on growing the world's most popular flower because they

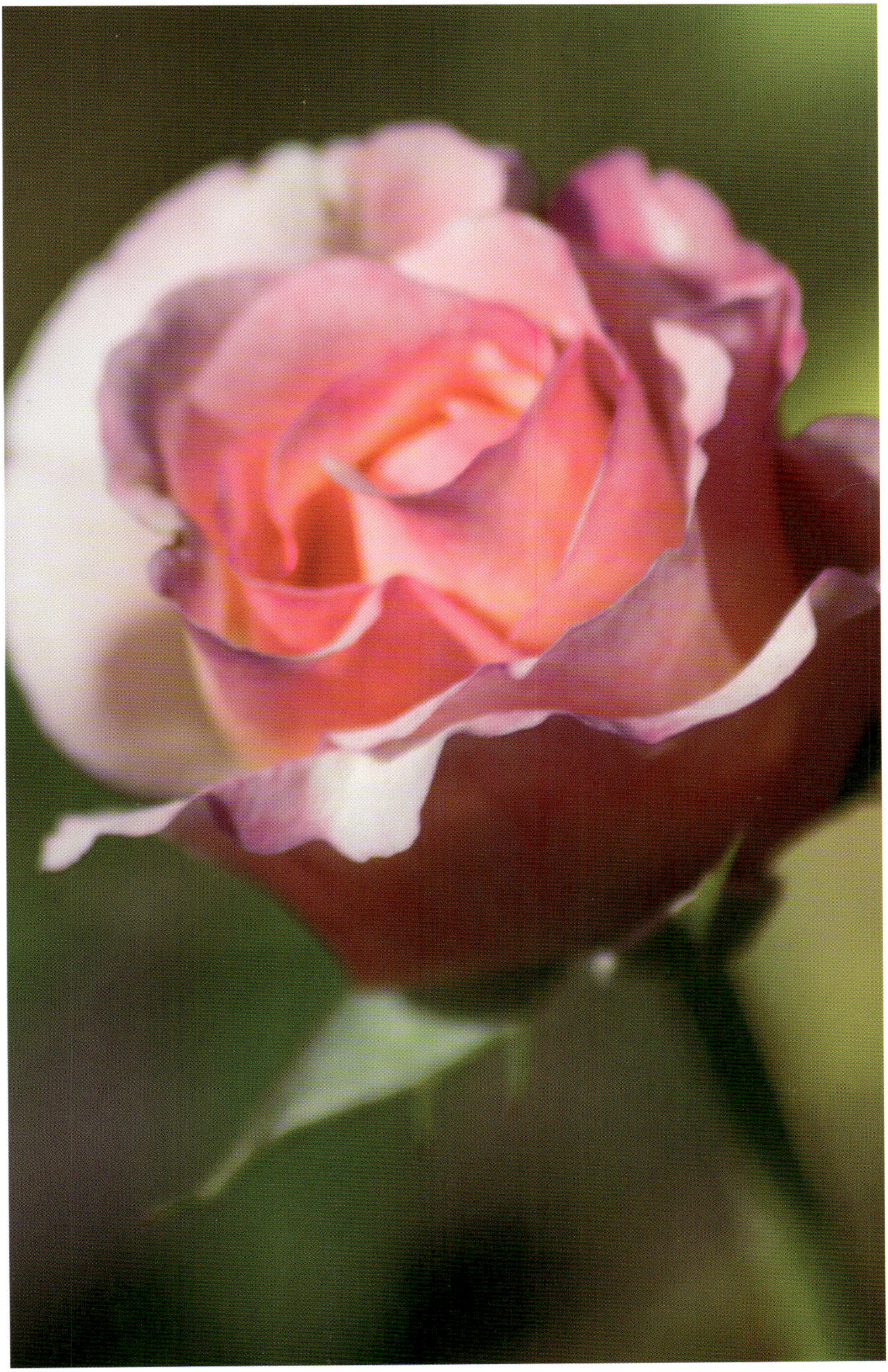

always die on you or bloom infrequently, give his technique a try and see what happens.

## First, dig a ditch

If there is any one secret he has, it is in the way he prepares the site before the roses go in. He digs a hole, a deep hole. Or if he is planting more than one rose, he digs a ditch, a deep ditch.

All roses require good drainage; those that stand in water for even a short time after a rain will flounder. Since most Kentucky gardeners are graced with a heavy clay soil that holds considerable water, roses should be planted in a way that allows their roots to stay above water. That's where the deep ditch comes in.

You need a stout shovel, spade or digging fork, and—if you have back trouble—a stout teenager who owes you something. Dig down the depth of the shovel or spade and remove the dirt. Then—and this is the tough part—dig down another shovel depth and remove that dirt too. You will need to do that about three times per plant. If the shovel or fork makes a fairly narrow trench, move over to make it twice as wide. You want to end up with a hole that is at least 2-foot-by-2-foot and at least two feet deep. Thirty inches deep would be better. Discard the old dirt; you're not going to be using it again.

In the bottom half of the hole, Dad puts bagged topsoil. You could also use gravel, composted manure or builder's sand. Builder's sand is coarse and will allow good drainage. What you don't want to put back in there is the heavy clay you just dug out.

With the bottom half full, it is time to put in the rose bush. I hope you have bought a No. 1 grade bush from a reputable garden center. It will have about a foot of root spanning out in spider web fashion. Clip off any damaged just above the knot where the hybrid was joined to its rootstock to protect it from freezing.

If you have used good soil or composted manure, the roses will need no fertilizer until after their first heavy bloom, usually in mid- to late June. Then you can put a cup of 10-10-10 fertilizer per plant and do that again in late summer before the fall bloom period.

Blackspot is the No. 1 disease threat to roses in Kentucky. It rarely kills the plants, but it leaves them looking ratty and reduces bloom production for the following year.

Dad sprays his roses once a week with a fungicide (Ortho's Funginex is one brand) to keep the bush free of blackspot. Making sure the plants are in full sun with good air circulation cuts down the incidence of blackspot. Also be sure to clean up the afflicted leaves (they will be yellow with black spots) and throw them in the garbage. Blackspot left unattended begets more blackspot next year. And a beautiful bloom on a sickly bush is a rather sad sight.

roots, splay the remaining root out in the hole and put the plant in so that the knot (if it is a hybrid) is at approximately ground level. Pour in a gallon or so of water, unless rain is imminent. Then steady the plant as you fill the hole with topsoil or composted manure. If a freeze or hard frost is still possible, pull the soil

## OUT, OUT, BLACKSPOT

Many gardeners give up on roses because of the persistence of blackspot. Blackspot does not harm the bloom but disfigures the foliage and leaves many plants nearly bare by midsummer. Regular—once a week in wet weather, less in dry—spraying of fungicides holds blackspot at bay. Another organic trick is to remove the bottom six inches of leaves on the plant. Blackspot typically is spread by spores in the grounds splashing onto the leaves. Take away the bottom leaves and you eliminate blackspot's toehold.

# Grow the signature shrubs of the South

In April and May, azaleas and rhododendrons will explode with bloom in the wild and in the spring landscape from Mobile Bay to the northern Appalachian Mountains. The shrubs bloom in colors from the purest white to rich lavender-purple and yellows, oranges, and reds in between. Unless the winter has been exceptionally harsh, azaleas and rhododendrons cover themselves in bloom. No shrub rivals azaleas and rhododendrons for show-stopping power.

Though they are differentiated in the plant trade, rhododendrons and azaleas are actually the same species—they are all rhododendrons. But we think of azaleas as generally shorter shrubs, with smaller leaves and more open flowers than rhododendrons. What we call rhododendrons are larger, more open evergreen shrubs with thick, leathery leaves. (From here on, I will call them both rhododendrons.)

Though they are natives, rhododendrons have a reputation for being difficult to grow. I know I have probably killed more of the rhododendron species than any single class of plants. That is because for years I planted them without paying attention to how they grow in nature.

Rhododendrons in the wild will almost always grow on a hillside near the edge of woods where they sometimes form dense thickets. They are almost never found in flatlands or in dense shade. They are far more common in the mountainous regions of eastern Kentucky than in the Bluegrass or western sections of the state.

If you can imitate how rhododendrons and azaleas grow in the wild, you have a good shot at getting the shrubs established. Here's how: Before you plant, pay attention to light, drainage, soil pH and variety.

## Light

Plant your roses in full sun, but save your rhododendrons for shadier spots. Those shrubs growing in the woods will get a filtered sun. Avoid, if at all possible, planting rhododendrons where they will get direct afternoon sun. They are more inclined to insect pests when in direct, hot sun, and their shallow root system will dry out quickly. Planting on the east or north side of the house or on the east side of trees or shrubs will give you the right light conditions.

## DRAINAGE

The reason rhododendrons are almost never found in flatlands or in heavy clay soil is because they require excellent drainage. They are subject to root rot if they stand in water, even for a short time, or if they are grown in heavy clay soil that drains slowly. Poor drainage likely kills more rhododendrons than every other problem put together, and unfortunately most of us garden in soils that are heavier and with more clay than rhododendrons prefer.

But that does not mean you can't have rhododendrons. One trick is to amend the heavy clay soil with a combination of two parts topsoil, one part peat moss and one part builder's sand. Builder's sand is coarse and allows for good drainage.

Another trick is to plant rhododendrons above the surface of the soil. To do that, take the plant out of its pot, set it on top of the ground instead of in a hole, and build up a mixture of topsoil and peat moss to the bottom of the plant. The key to success with this method is to be sure to keep the plants watered for the first summer, especially if the summer is hot and dry. As mentioned earlier, rhododendrons have shallow roots and will dry out quickly.

## SOIL PH

I think more is made of the importance of soil pH to rhododendrons than is really necessary; good drainage is far more important. Still, in their native regions, rhododendrons grow in a soil that is more acidic than that found in most Kentucky gardens. If you are unsure about your soil's acidity or alkalinity, have your local Extension Service office run a test for you. If the soil has a pH of 6.5 or higher, add a cup of a soil acidifier, such as sulfur or aluminum sulfate, when you plant and add a cup per plant every spring thereafter. Some fertilizers are made especially for rhododendrons and azaleas and acidify the soil as they add nutrients.

### VARIETIES

Most varieties of azaleas and rhododendrons available to home gardeners are hybrids. There are literally hundreds of hybrids available, and gardeners can be bewildered by the array.

For cold-hardy rhododendrons, *The Southern Living Garden Book* recommends 'America' (dark red), 'PJM' (lavender), 'Ramapo' (pinkish violet) and 'Nova Zembla' (red).

Rhododendrons can grow to huge proportions and get out of control. But there are low-growing varieties. Try 'Blue Diamond' (lavender blue), 'Molly Ann' (pink) and 'Elizabeth' (red).

Azaleas come in evergreen and deciduous types. Among the deciduous types, the Exbury hybrids are popular. In the evergreen types, the Monrovia varieties usually grow very well here.

You might also want to try the native species rhododendrons, especially if you like more naturalized landscapes or if you have had bad luck with the hybrids. Sweet azalea is a hardy shrub that will grow up to 20 feet tall. Flame azalea is a native that extends the bloom period into mid-summer. Swamp azalea is the one rhododendron that can tolerate poor drainage and is native to the mid- and upper South.

Rhododendrons' burst of color in the spring is worth giving them a try. And if you give them what they want and they still won't thrive, quit trying and put in a viburnum or some other tough shrub. You don't want to become a serial rhododendron killer.

## EVERBLOOMERS!

I haven't tried them, since I kill most azaleas and rhododendrons I adopt anyway, but plant breeders are introducing more and more plants that are "everblooming," including azaleas. That has to be a temptation to gardeners, because azaleas in bloom are certainly magnificent and much more mundane when not clothed in their garish colors. These "everblooming" plants typically bloom heavily in spring and sporadically after that. I recommend trying them on an experimental basis, but don't make them a keystone in your garden until you see how they perform.

# Royal purple in the garden

Once upon a time, we common folk could not wear purple. It was the color reserved for royalty. It is just as well. The only source of purple in ancient times was a snail that was harvested off the coast of present-day Israel. It took hundreds of snails and great labor to produce a purple cloak that distinguished a Roman noble from the riffraff, so garments of that color were both rare and expensive. The common folk couldn't afford purple garments even if they could find shoes to match.

Happily, gardens are more democratic. Purple is a richly royal color, but it adorns a range of plants from trees to grasses, perennials to spring bulbs, annuals to shrubs. Purple is an almost-red and maroon on some varieties of clematis, viburnum, allium and oakleaf hydrangea. Purple is nearly white to pale lavender on some tulips, wisteria and roses. Purple is deep and pure on some varieties of German iris, Siberian iris, salvia, catmint, clematis, echinops and petunias.

With a little planning, something purple can be blooming in the garden all year long.

In early spring purple is on the crocus, the grape hyacinth, the Dutch iris and, of course, the tulip. One of the finest tulips ever created is a Darwin hybrid called 'Shirley,' which is white with each petal feathered in purple. 'Queen of Night' is a purple tulip that is almost black; 'Arabian Nights' is a deep, rich purple, the color of a Roman senator's robes. And much less pretentious.

My favorite early-blooming shrub is the lilac. Breeders have produced white and yellow and variegated lilac flowers on the plants, but the lilac, or light purple color, on the old-fashioned shrub that your grandmother knew is still my lilac of choice, both for color and fragrance.

If you have shade and want purple in early spring, consider the hellebore, which blooms in rich shades, including deep purple.

Mid-spring brings on my favorite source of purple flowers, the iris. The iris that we called flags in the old days were a lavender color; some were two-toned with light purple standards, or upright petals and deeper purple falls, or lower petals. The iris variety 'Ace of Spades' unfurls jet black but becomes deep purple when it opens. Breeders have created other purple iris so rich in shade they would make a king cry.

Mid-spring also brings the arrival of the first salvias, some of which, including 'May Night' are a dark purple. Salvias combine well with the purple iris both in flower color and plant form.

Wisteria can be a nuisance vine with its tendency to overgrow everything in sight and pull gutters away from the siding. But a wisteria in bloom, its purple chains of blossoms dripping from the branches, is almost magical. Another late spring favorite is 'Jackmanii' clematis. Breeders have produced clematis with larger flowers than this original hybrid clematis, but they have yet to match the jackmanii's ability to cover a fence or mailbox in the most royal shade of purple imaginable.

In shade, the columbine 'Magpie' offers a happy combination of deep purple and white with just a hint of yellow.

Lavender, phlox, catmint, Russian sage, echinops, lobelia and purple coneflower are all excellent sources of purple in hardy perennials that will bloom unceasingly in late spring and during the heat of summer. Annual sage and petunias are good choices for getting a dash of purple in planters on the deck or porch.

Some plants have varieties with purple foliage rather than flowers. Heuchera is an example. Coleus is another, and it likes the shade.

Plant 'Black Knight' butterfly bush and get purple blooms and butterfly visitors all summer. 'Purple Majesty' millet is an All-America selection of grass equally striking in a container or border.

In the fall, mums the word for purple possibilities. The foliage of many shrubs and trees also turns purple with the coming of shorter days. Some varieties of viburnum change to deep purple-maroon. So does oakleaf hydrangea. Breeders have created several varieties of ash trees with leaves that turn purple on top and buttery-yellow on the bottom—striking on a sunny fall day.

## Planning for purple

The garden writer Gertrude Jekyll didn't like the color purple and almost never used it in her designs. "Malignant magenta," she called it.

Well, good for her. Jekyll liked planning large sweeps of borders for her gardeners to take care of. And it is true that purple does not show well in sweeps from the castle heights. But down close to the earth, where most of us work our own plantings, purple looks good up

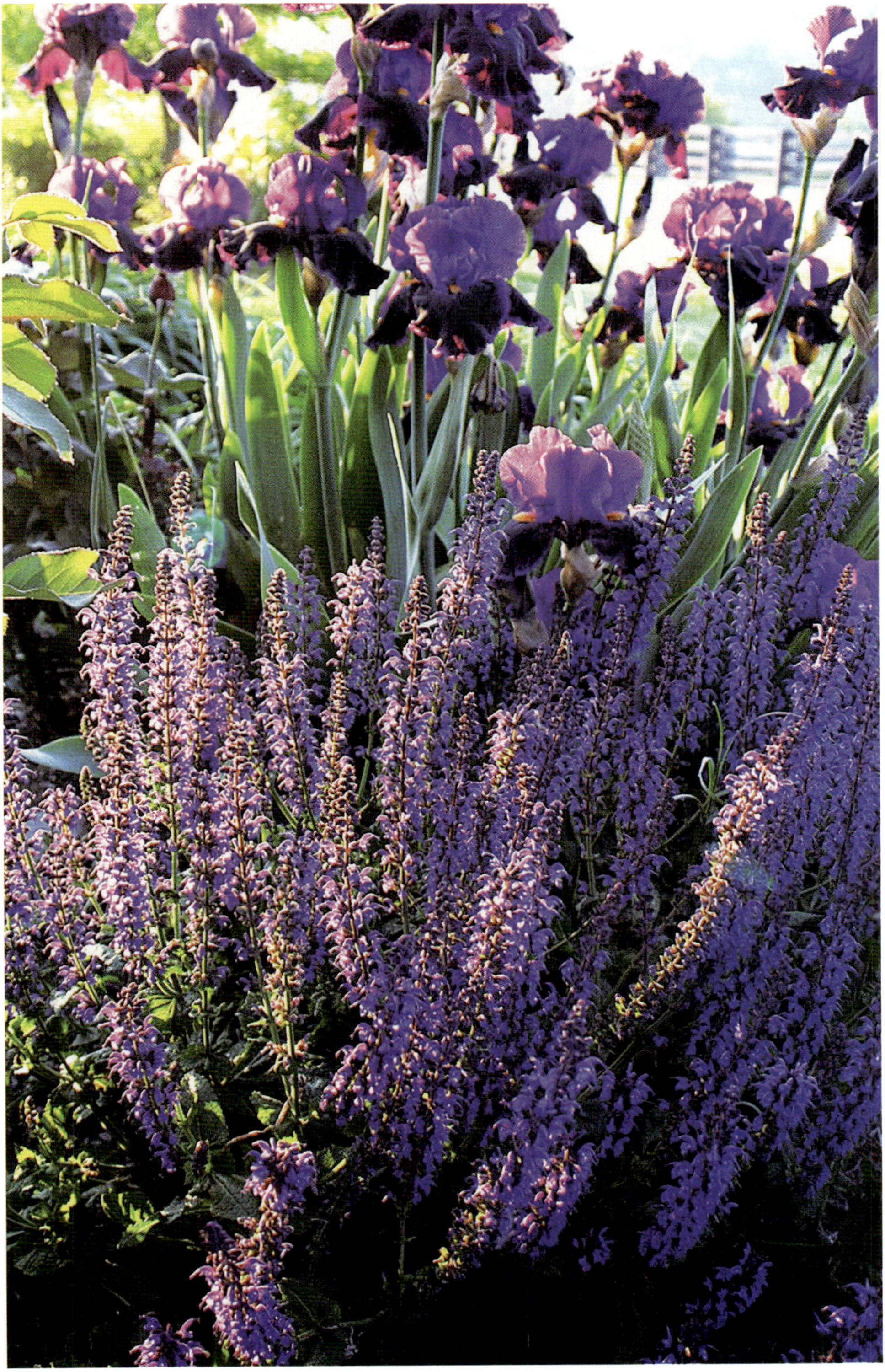

Or those same black-eyed Susans could stand in front of a purple butterfly bush. Ditto purple salvia and achillea 'Coronation Gold.' Or white and purple Siberian or German bearded iris for springtime charm.

One of the finest combinations of purple and white for summer bloom I have ever seen has been 'Munstead' dwarf strain of lavender fronting white 'Iceberg' roses. Plant them along a walk and you have a vision worth a king's ransom.

close and personal.

The design folks say the best color to match with purple is its opposite on the color wheel: yellow. Combine light and lavender purples with deep yellows. Combine deep purples with paler yellows. And white and purple are always correct. Or so they say.

So an easygoing, heat-of-the-summer combination would be purple coneflower paired with yellow black-eyed Susans.

## ONE 'PURPLE' TREE TO AVOID

I get asked the question, usually in spring: "What's that big tree blooming with purple blossoms?" You could be seeing purple wisteria climbing up the branches, appearing to be part of the tree. More likely, you are seeing the blooms of a tree called paulownia, or empress tree. Pretty in the spring, maybe, but you don't want it. It grows so fast it's scary, and its wood is weak, so it's likely to come crashing down someday, purple blossoms and all, on something you hold near and dear.

# Back to the garden

A recently released survey by the Independent Garden Center Association advised its members to get ready for a resurging interest in vegetable gardening. The survey predicts more and more gardeners are looking to spend their hard-earned green this year on plants, seeds and supplies that will put some food on the table.

Four trends are driving a renewed interest in food gardening: the state of the economy, food safety worries, concerns about the environment and a growing interest in cooking with fresh ingredients.

A carefully planned, well-maintained vegetable garden can easily yield the equivalent of $4-$5 in food cost savings per square foot. That means a homeowner with a very small patch, say 10-foot-by-10-foot, could save $400 to $500 per year. Most of us have room for much larger gardens.

Food poisoning continues to make news. You are highly unlikely to eat tainted food if you grow it yourself. The salmonella and E. coli contaminations in produce came from farms where food was produced in high volume in the U.S. and foreign countries. You are more likely to win the lottery than get food poisoning from vegetables and fruits you grow yourself.

Not only does long-distance food come to you at slightly higher risk of contamination, it comes at a high environmental price. The average morsel of food consumed in this country travels 1,400 miles before it reaches your plate. It uses more energy to get there than the calories it produces. Walking out to your backyard to find food rather than buying something off the shelf that was grown in Chile or China is a savings to both your pocketbook and the environment.

Finally, perhaps thanks to the Food

Network, more and more consumers are interested in cooking with fresh, locally grown ingredients. A tomato doesn't get any fresher than the one picked in the morning and served for lunch that same day.

## The basics

Starting a productive vegetable garden means paying attention to what I call the three S's—sun, soil and size.

It is absolutely critical that a vegetable garden get at least six hours of sun every day, and eight is better. The greens—lettuces, kale, etc.—can get by with a little less than six hours of sun, but even they are happier with more.

I would never advocate cutting down trees if you live on a shady lot, but you may be able to limb them up to gain more sunlight. But have a pro do that or you may irretrievably damage the tree. If all you have is shade and can't remove limbs, you may want to rent a garden spot or offer to help a neighbor or friend in the garden for a share of the produce.

## Soil is the next critical factor.

Good garden soil—a fertile loam—is hard to come by in most of the state. Most soils in Kentucky are either sticky clays or infertile, sandy soil.

But all soils can be improved with the continued application of organic materials (leaves, straw, grass clippings, etc.) and manures. Before planting, take a soil sample to your local Extension Service office and let the agency determine what your soil needs to grow a good vegetable garden. Then add amendments according to the directions given. The adage is: Feed the soil and it will feed you.

One soil situation you absolutely want to avoid is poor drainage. No vegetable plant will tolerate long periods of standing water. So do not site a vegetable garden in a spot where water stands for half a day or longer after a heavy rain. The best spot for a vegetable garden is on a gentle slope in full sun.

## Finally, size

A vegetable garden is labor-intensive. Too many gardeners start a vegetable patch only to abandon it by mid-August to the heat, bugs and weeds. Do not plant a patch so large you can't control the weeds, keep the vegetables insect-free or

## GET THE HIGHEST RETURN FROM YOUR PATCH

If your vegetable garden space is limited and you have set your sights on saving the most on your grocery bills, you may want to consider limiting your planting to the crops that have the highest dollar value. Below is a list of vegetables you can grow that have a high, medium or low dollar value:

**High:** lettuces and greens, green onions and leeks, tomatoes, peppers, asparagus, okra, strawberries, blueberries, raspberries.

**Medium:** melons, green beans, peas, summer squash, carrots, broccoli, cauliflower, blackberries, storage onions, herbs.

**Low:** dried beans, sweet corn, Irish potatoes, sweet potatoes, winter squash, cabbage, turnips.

harvest the crops in a timely manner.

To avoid the order-and-plant-everything-in-the-catalog syndrome, make a list of the fruits and vegetables your family will eat, or at least try. Then cross off the list anything that is too much trouble or too inefficient to grow (see box). For example, your family may love celery sticks and peanut butter, but celery is very difficult to grow in Kentucky. So are peanuts, for that matter. Then match your list to your space, remembering that you can rotate vegetables in and out of the garden depending on the season. For example, peas can be planted in March. When they are finished producing, about mid-May, you can put green beans or summer squash in their place. Then when the beans or squash expire—say mid-August—add some greens, broccoli or cauliflower. That's what is called maximizing your assets.

To help you plan your space, I highly recommend the booklet "Home Vegetable Gardening in Kentucky" available at your local Extension Service office. It's not only chock-full of information, it's free.

## PLANT WITH YOUR FEET

Here's a trick for planting Irish potatoes: Use your feet. To get good-sized spuds, you need to plant the seed pieces at least a foot apart. You also need to make sure the pieces get good soil contact. What I do is dig out a trench in loose soil a few inches deep and walk in the trench, dropping the pieces as I go. I step on a piece with my heel and drop another piece at my foot. Then go back and cover the trench with several inches of soil. You want to keep the soil covering the potatoes as they grow; exposed spuds turn green and bitter, poisonous actually. But surely you're not eating the skin!

# Tips to make your garden greener

Our gardens should be things of beauty and oases in a sea of asphalt, concrete and brick. They should nurture our bodies and our spirits and offer refuge from stress. Each person's garden should be an individual contribution to making a piece of the world greener, quieter and more natural than the rest of the mess we call progress and civilization.

Unfortunately, we Americans too often do to gardening what we have done to farming—industrialize it. We overuse pesticides, insecticides and fertilizers; we let our lawns sprawl over vast acreages and then attack them with heavy-duty machinery; and we plant exotic specimens that need a whole shelf of garden-center products for their care.

In short, our gardens and landscapes, which should contribute to a world that is "greener"—in all senses of that word—are all too often biologically sterile and visually and spiritually unsatisfying.

Following are some tips for creating a "greener" garden.

## Grow your own food

The single best thing you can do to strike a blow for a greener tomorrow is to grow as much of the food you eat as possible.

The average bite of food on the American plate travels about 1,500 miles and has used far more calories of energy getting there than it will put into your body. That is because we insist on eating any food at any time of the year, and we will buy only food that is picture perfect, blemish-free. So it comes to us from California or Chile or Florida.

If you have never grown your own food before, start small, perhaps with a few tomato and pepper plants. These can be grown even on the balcony of an apartment building. If space allows, try beans and squash. Once successful with those, branch out and try broccoli, cauliflower, cabbage and melons.

Kentucky author Barbara Kingsolver wrote about her family's experiment with eating only from their own garden or what they could buy locally in *Animal, Vegetable, Miracle: A Year of Food Life*. Kingsolver's family managed to grow

SPRING 31

or buy from local farmers more than 90 percent of the food they consumed for a year. Depending upon the size of your space, you may not be able to achieve that, but it is a worthy goal.

## Shrink your lawn

Depending upon its age and repair, the average lawn mower sends anywhere from 20 to 200 times more pollution into the air than an automobile. Cutting back on mowing may be the next best thing you can do to improve our environment. The best way to do that is to minimize the size of your lawn.

I have a friend who has replaced all of the grass in her back yard with shrubs, perennials, a water garden and a vegetable garden. Paths snake through the plantings; there is no grass to be mowed.

Depending upon the size of your lawn, that may not be practical, but you can whittle away at the size of your mowing area by adding perennial and shrub beds. If designed for easy maintenance, raised planting beds can be nearly carefree, especially after they are well-established. Shrub beds are probably easier to maintain than a mix of perennials, and they are also more likely to attract birds.

Whatever you do, avoid the "wildflower-in-a-can" mixtures that promise to replace your lawn with a beautiful carpet of wildflowers. Won't happen. What you will end up with is a weedy mess that ticks off the neighbors and eventually needs heavy-duty machinery to mow down.

## Cut back on fertilizers, pesticides and herbicides

It doesn't take anything more than a trip to a discount store's garden center to see that we in America douse our gardens with large doses of chemicals to kill weeds, insects and disease. We are especially profligate with the chemicals we pour on our lawns in search of the picture-perfect sea of green.

With careful choices of grasses, perennials, shrubs, trees and vegetables, we could eliminate many chemicals and cut back on the use of others. For example, many varieties of roses, especially hybrid tea roses, are very vulnerable to a disease called blackspot and require persistent spraying with fungicides to keep them handsome. But many roses—the 'Knockout' series, for example—resist blackspot and need no spraying. Even among hybrid tea roses, varieties vary a great deal in their resistance to disease.

My rule of thumb is that if you have a plant that needs an annoying amount of coddling with chemicals, get rid of it and try something else. And never, never plant a species in the same spot where one of its kind died previously.

## Go native

This follows from the advice above. Native annuals, perennials, shrubs and trees have evolved in this climate with local pests and have learned to survive. Native plants have withstood our heat, our droughts, our humidity, our late freezes. So in your garden, they are likely to be hardier than species whose homeland is China, Japan or Africa. My friend who has eliminated her backyard lawn has planted nearly all native specimens. Shooting Star Nursery in Georgetown is a good source of native plants.

## Plant a tree every season

Trees shade us from summer sun and block wintry winds; they cleanse the air; they provide habitat for wildlife. Just as in World War II when we were encouraged to plant Victory Gardens to help the war effort, today's Americans should plant trees—at least one in the fall and one in the spring—to combat global warming. If you don't have space to plant a tree, donate one to a community park, arboretum or forest.

## SOAP 'EM

When you must spray fungicides or insecticides in the garden, add a couple of drops of a mild dishwashing soap to the spray water. The soap helps the water stick to the plants' leaves. Read the label, though. Some sprays should not be mixed with soap, and some sprays already contain "sticking" agents.

# Do justice to your roses—Monty's way

If you want your rose bed to be a blanket of roses this summer, try some of the late Monty Justice's advice. Because Monty knew roses.

He raised roses for 40-some years. He was president of the Louisville Rose Society, raised hundreds of varieties of roses. He cared for about 100 private and public rose gardens, including the rose beds at the Kentucky Fair and Exposition Center. He developed a series of rose plant food called Monty's Joy Juice.

Keep that in mind because some of Monty's advice on growing roses is unorthodox and may contradict everything you have ever thought you knew about growing roses, or any other kind of plant for that matter. But if you ever toured Monty's rose gardens and admired the lush foliage and perfect blooms of his hybrid teas, grandifloras and floribunda roses, you know you better heed his methods.

Here's the Monty Justice way of growing roses:

Start with the rose you bring home from the garden center. If it is a rose wrapped in cardboard, soak the plant for one to two days in water before you put it in the ground. If the rose came in a pot, you can skip that step.

Choose a site in full sun. Dig a hole 10 to 12 inches wide and 15 to 18 inches deep. Throw away all of the soil that came from the hole. To the bottom of the hole add any porous material— perlite, sawdust, manure or sand. Then place the plant in the hole so that the bud union (a swollen place where the stems join the base) is just below ground level. Keeping that bud union underground makes it unlikely the rose plant will succumb to winter kill, Justice believed. Then fill the hole with a good potting soil. Justice recommended fertilizing with orange label Monty's Joy Juice mixed with water. The fertilizer is high in potassium, which protects the plant from insect and disease attacks, he said. Fertilizer, whether liquid or granule, should be placed near the base of the plant and watered in.

The usual garden advice is to water plants infrequently but deeply. Justice said do the opposite with roses. Water frequently. Water moving down into the root system brings much-needed oxygen to the roots and removes carbon dioxide, allowing the roots not only to drink but to breathe better.

"It's not how much you water, it's the frequency of watering," Justice said.

Roses will typically have their first flush of blooms in late May to early June. Justice said rose growers need to resist the urge to run out and cut armsful of roses to bring indoors during that first bloom period. Instead, the roses should be left on the plants and the bloom removed only after it has started to wither. That practice will keep the plant bushy and compact; removing rose stems will make the plant leggy and tall, he said.

Also after that first bloom period, gardeners need to remove the leaves from the bottom six inches of the plant. It is those leaves that are most likely to get and spread that most dreaded of rose diseases, black spot.

Besides removing the bottom leaves,

---

When you have grown more 400 varieties of roses, it's hard to pick favorites, but these were some of Justice's favorites for Kentucky Gardens:

**Red:** 'Veterans Honor'
**White:** 'Elina or Crystalline'
**Red and white:** 'Double Delight'
**Yellow:** 'Gold Medal'
**Light Pink:** 'Bride's Dream'
**Lavender:** 'Fragrant Plum'
**Orange Pink:** 'Touch of Class'
**Orange Red:** 'Fragrant Cloud'

---

## GIVE 'PEACE' A CHANCE

The 'Peace' rose is still the most popular hybrid tea ever created. Bred in France just after World War II, 'Peace' got its name from the hope that the horrors of world war were over forever. Yellow, with margins of pink, 'Peace' also is one of the most beautiful roses ever bred. Modern breeders have far surpassed 'Peace' in hardiness and disease resistance, but they haven't eclipsed its beauty or popularity. Every rose garden needs 'Peace.'

gardeners should spray roses once a week to combat blackspot. Justice developed an organic spray that he claimed works just as well as the chemical sprays for the disease. Here is the recipe to make one gallon of spray: one-half tablespoon of Monty's Joy Juice, one tablespoon of Epsom salts, one tablespoon of baking soda and two and a half tablespoons of light vegetable or canola oil. The baking soda in the mix fights disease and the oil suffocates insects. The mixture should be agitated well to mix the oil thoroughly. Spray the undersides of the leaves because that is where black spot gets a foothold, Justice said. For larger rose plantings, Justice recommended spraying once a week with Daconil Ultrex.

As for winter care, Justice said if the rose was planted deep enough to cover the bud union and kept fertilized with a high-potassium fertilizer, the plant should survive the winter well.

Another tip from the late rosarian: if the gardener has to move a rose during fall or winter, the rose should be dug up, the dirt shaken loose from the roots and the plant stored in a cool place inside of a garbage bag. The garbage bag will keep the plant's roots moist.

To get more of advice on rose growing or to buy a quart or two of Monty's Joy Juice, you can access the company's website, www.mymontys.com.

# Try something new

March is an excellent month for finalizing those garden plans for the year. Though it is a good time to plant trees and shrubs, most vegetables, annuals and perennials won't go in until April, when the real work of the gardening year begins.

I'm a pretty conservative person and an even more conservative gardener. I like to stick to what works. Gardening is too much work to waste time getting lured by Madison Avenue marketing slogans for plants that sound good on the page but droop in the garden. Nevertheless, every year I add a few new species and varieties to my "must buy" list just to see what will happen. Sometimes I am pleasantly surprised. For example, for years I planted 'Silver Queen' corn and wouldn't even consider changing. Then I tried a variety called 'Silverado,' and yes, it's better. The flavor is about the same, but 'Silverado' produces more ears and resists a creepy disease called smut.

Each year, literally thousands of new plant varieties and garden products appear on shelves in local garden centers and in the pages of seed catalogs. These are a few recently introduced items you may want to try. I always recommend you search for them first in local garden centers, and then try catalogs or do an Internet search on the plant if you can't find them there.

## Shrubs

Always tread carefully when you buy any untried type of shrub because you are committing yourself to several years of growth before you know whether you have a winner.

The lilac is one of my favorite shrubs both for blooms and the heavenly scent. But lilacs certainly have a short bloom period and it's early enough in spring that sometimes the blooms get blasted by the cold. So the idea of a reblooming lilac is appealing. One variety to try is 'Josee,' which has pink flowers and will rebloom in the summer after the spring burst of color.

Lacecap hydrangeas are becoming deservedly popular both for their summer blooms as well as for the hardiness of the plant itself. Lacecap hydrangeas, like lilacs, generally put on a burst of bloom and are done. But the 'Twist-n-Shout' lacecap hydrangea is heralded as a rebloomer. Another hydrangea that looks delicious is 'Vanilla Strawberry,' which is likely available locally.

## FLOWERS

The array of new flower varieties every year can be bewildering, especially to the new gardener. One guide may be the selection of plants that have won awards from national organizations.

Flowers and vegetables that have earned the All-America Selections award for the year are likely to be decent performers in your garden, though I have, on occasion, been disappointed. These plants have thrived in trials conducted in several parts of the country.

Recent winners are 'Mesa Yellow' gaillardia, 'Endurio Sky Blue Martien' (Who names these things?) viola, 'Twinny Peach' snapdragon, and 'Zahara Starlight Rose' zinnia.

Gaillardias are tough flowers that love the heat of the summer, and 'Mesa Yellow,' a plant with controlled habit and 3-inch blooms, sounds like a good one. It is smaller than most gaillardias, reaching 20-22 inches in full sun. Snapdragons can get lost in the garden, but 'Twinny Peach,' with its double blooms should put on more show. Snapdragons make a good cut flower. Violas look delicate but the plants are tough, and 'Endurio Sky Blue Martien' promises a burst of blue in the spring. Zinnias fell out of favor for a while because they were prone to powdery mildew. But breeders in the last decade produced dozens of cultivars that stay clean. 'Zahara Starlight Rose' is the first rose and white bicolor zinnia that also carries disease resistance. Its 12- to 14-inch height makes it perfect for the front of the summer sun border.

All-America Selections plants and seeds should be widely available.

Other flowers for your consideration include a double form of the popular Knockout series of roses and 'First Frost' hosta, the 2010 Hosta of the Year winner from the American Hosta Society.

## VEGETABLES

Old is new and new is old in the vegetable garden as heirlooms continue to be discovered and offered to the public. While heirloom vegetable varieties, especially tomatoes, continue to appear on local garden center shelves, the best mail-order source for heirlooms is Baker Creek Heirloom Seeds. "New" heirlooms include the 'Scotch Bonnet Yellow' pepper that is so hot it comes with a warning to "use with caution." Another pepper, 'Yellow Monster,' is sweet and huge—8 inches long by 4 inches wide. 'Gypsy' tomato is touted as the deepest purple.

One of the favorite heirloom vegetables of all time is the 'Brandywine' tomato, regarded by many as the best flavored in the world. But it's a tough customer to grow, at least here in Kentucky. 'Brandy Boy' tomato is a descendant of 'Brandywine' but supposedly easier to grow. We'll see.

## Garden Stuff

As gardeners try to get away from using chemical sprays to deter insects and disease, organic products have proliferated. Captain Jack's Deadbug Brew Flower and Vegetable Garden Dust, promises to kill bugs without poisons. Another organic product, Hot Pepper Wax, can be used to kill insects inside and outside. Its key ingredient is capsaicin, the stuff that makes hot peppers hot. I've certainly tried some peppers that would kill lesser beings. On the other hand, my experience with organic products in the past has been mixed; they don't always perform as well as claimed. But who wants poisons on food you will eat?

Going greener? Look for Earth Friendly Expanding potting mix in local garden centers. The mix is made with worm castings, among other organic ingredients.

If you're starting your own seeds this year, remember that most indoor seed-starting projects fail because of a lack of light. Consider getting one of the seedling light setups that includes fixtures with bulbs that provide the "sunlight" your fledgling plants need for a good start. Costs will run anywhere between $100 and $1,000, but over time the light stands will pay for themselves because it's cheaper to start your own seeds than to buy plants. Or you can go conservative and make your own plant stand out of scrap wood and a shop light. Works just as well.

## ANOTHER OLDIE BUT GOODIE

While plant breeders continue their efforts to improve home garden varieties, some of the oldies perform as well, if not better, than new models. One of my favorite watermelons is the heirloom 'Dixie Queen.' It is a round melon, light green with dark stripes and in good years can weigh 20 pounds or more. It is, by far, the best watermelon I have ever eaten, and better still, it is one of the most dependable. Hard to find but worth the effort; try Internet searches of heirloom seed houses.

# In love with lilacs

When my family and I lived in Dayton, Ohio, we were fortunate enough to have bought a house where we inherited an entire row of lilacs that bordered our yard and the neighbors' property. Every spring, the shrubs covered themselves in the cone-shaped blooms that filled the springtime air with wonderful fragrance. To my nose, nothing in nature smells better on the spring air than the sweet scent of lilac.

Upon moving back to Kentucky, we found a great little farm in a wonderful location, but the previous owner's idea of landscaping was to plant a five-acre sweep of Kentucky 31 fescue. No shrubs, few trees and certainly no lilacs. To help with the landscaping—and the budget—my mom decided to dig out "a piece" of lilac from an old homestead near her house and give it to me. Her effort earned her a severe case of poison ivy, but I got a shrub that bloomed next to my driveway for years, offering passersby its sweet promise of spring for a few weeks in April. When we left that house, I dug up a piece for our new place. But perhaps because I didn't acquire a case of poison ivy in the taking, my scion from mom's plant pouted for a few years, and then died.

That lilac was the Syringa vulgaris (common lilac) species, a shrub that grows 8- to 20-feet high. It bloomed "lilac" purple in the spring and, unfortunately, covered itself with powdery mildew every summer so the leaves looked like someone had spilled milk on them.

Today, lilac lovers can find Syringa vulgaris varieties that are bred to resist the mildew problem and that also come in smaller forms to fit smaller backyards, even patios. And "lilac" isn't the only color available anymore. You can find lilacs in a range of colors from white to yellow to pink, red, blue and on around the color wheel to deep purple. 'Dwight D. Eisenhower' is an outstanding blue variety while 'President Grevy' is an excellent double-flowered blue. 'Angel White' is an attractive white-flower lilac. Many of the new and improved lilacs are in the purple class; 'Flower City' is an almost pure purple, and 'Agincourt Beauty' is a deep purple lilac. If the old "lilac" color is still your favorite, as it is mine, you may want to search for a variety called 'Mollie Ann,' a lilac with large, showy heads that was introduced by Catholic priest and noted lilac hybridizer Fr. John Fiala in 1981. Perhaps the most stunning lilac, certainly for color, is 'Sensation.' It's a deep violet with each petal edged very daintily with white. Plant 'Sensation' in a spot where you will see it close up to appreciate the color of the petals.

Local garden centers are carrying more of the Asian or "Chinese" or "Korean" lilacs, and they are becoming more common in Kentucky landscapes. The Asian lilacs are less prone to powdery mildew than the Syringa vulgaris species. They also tend to bloom a little later—early summer—than the common lilacs, so they avoid frost damage to the blooms. Many cultivars of Asian lilacs, including the popular 'Miss Kim,' cover themselves in bloom in late spring and early summer in north-central Kentucky. Some of these species have excellent fall color and a few have variegated leaves, but they tend to be less hardy. The downside to me is that the Asian lilacs just don't have the same light, sweet fragrance the common lilacs have. They are sometimes fragrant, to be sure, but the smell is almost overpowering. And many have no smell at all. Lilacs without the smell is like potato chips without salt.

## Culture

In Kentucky, lilacs are near the southern edge of their comfort zone. They are a shrub better suited to the colder climes of the upper Midwest, in part because of that pesky powdery mildew problem they encounter in the humid Southeast.

But they certainly will grow and thrive here, once well-established. Good drainage is the key; choose a site for your lilac in full sun in a spot that never holds water for more than a few hours after the heaviest of rains. Unfortunately, most of Kentucky's soils, with their heavy clay content, are poorly drained. So lilacs would be best planted on a slope.

Another option is to build a layer cake of loose soil following Fr. Fiala's recipe in his book *Lilacs: A Gardener's Encyclopedia*. First, lay down 6 to 8 inches of gravelly sand directly on the surface of the plant site. Then add 4 to 6 inches of well-rotted manure or humus. Next add 6 inches of good topsoil. Place your lilac on the topsoil and put another foot of loose soil over the roots. You thus have created a little slope that will keep your lilac's roots out of standing water. (The same technique also works well for roses, azaleas and rhododendrons.) A lot of trouble, but remember that this plant, if it gets off to the right start, will outlive you. So it's worth a few extra minutes of your time at planting.

Do not heavily fertilize lilacs. A side-dressing of compost or bagged manure once every year just after blooming is enough to keep the plant growing well. A few handfuls of limestone around the base won't hurt, either, especially if you live where soils are acidic.

Since lilacs bloom on new wood, you should prune out about one-fourth to one-third of the oldest branches as well as the suckers every year. If you want to keep the blooms at eye level, prune back the canes to 8 or 9 feet.

Finally, you can start new lilac shrubs by digging up the suckers that spring up around the outside of the plant. Just watch out for poison ivy.

## REBLOOMERS

Just as they have with azaleas and hydrangeas, plant breeders are introducing more and more lilacs that rebloom. I haven't tried them and would recommend doing so only experimentally. While breeders have done wonders developing new colors and creating plants that avoid powdery mildew, they haven't much improved on the combination of wonderful lilac color and smell of the common lilac that has graced homesteads for generations. Try the new, but don't throw out the old just yet.

# This Judas won't betray you

Every year, in early to mid-April, I find myself driving down I-65 toward the Mammoth Cave area. Just north of Elizabethtown, along a shelf of almost bare rock high above the interstate, stands a line of redbud trees. If the early morning sun hits the trees, their purple-red blossoms sparkle. If the morning is overcast, the redbuds light up the gloom. Either way, it's a sure sign to me that spring, and, finally, warm weather are on the way.

## The Judas tree

The redbud (Cercis canadensis) is frequently called "the Judas tree" because, according to legend, Judas Iscariot hanged himself on a redbud tree after he betrayed Christ. The tree's blossoms, originally white, turned pinkish red after Judas hanged himself out of (take your pick) shame or because of his blood.

A creepy story attached to a really fine little tree.

Whatever its past, the redbud is a native, small, flowering tree that has been overlooked by gardeners, who are more likely to spend their spring gardening bucks on the gaudier flowering dogwood. But in recent years, gardeners and plant breeders alike have started to warm up to the redbud, a tree with just as much of the charm but with fewer "medical issues" associated with dogwoods. (A nurseryman told me once that homeowners planting dogwoods better plan to lose about 50 percent of those trees!) Unlike the flowering dogwoods, redbuds transplant easily, thrive in a wide range of soil types, grow quickly and have few insect and disease problems. And while the native species found growing in the woods is a perfectly good tree, plant breeders, seizing on the growing interest in redbuds, have developed a dozen or so cultivars with blossoms of different colors, double-flowered blooms, richer fall color, and weeping and dwarf forms.

## In the landscape

A member of the pea family, the redbud typically grows 20- to 30-feet high and can spread 25 to 35 feet. The redbud usually is a multi-trunk tree, splitting off close to the ground, though it can be pruned to a single stem. The redbud flowers in early to mid-spring, sometimes when snow is still on the ground. Individual blossoms are tiny, but they come in clusters that make for a fine show. The redbud's blooms cling to branches and even sprout in clusters on the trunk of the tree. Because the blooms hug the tree, they are able to withstand colder temperatures that would zap the showier blooms of the dogwood.

Some see the buds of redbud as red, some see pink and others purple. To my eye, the blooms are a combination of all three colors most of the time, depending upon the light.

The redbud's leaves are also attractive—large and heart-shaped. A subspecies of redbud, Cercis texensis, common in Texas and Oklahoma but able to grow here, has shiny leaves. In the fall, the redbud's leaves turn a buttery yellow. One cultivar, 'Hearts of Gold,' was bred especially for its spectacular yellow fall foliage.

## RED (AND WHITE) BUD CULTIVARS

Plant breeders have been working overtime creating or discovering redbud cultivars, though for my money the species works just fine. If I did plant a cultivar, I would go with 'Appalachian Red' for its striking blossom color that is closer to scarlet than the species. (I have never seen the white cultivars, but a redbud with white blooms just seems wrong.) Here are some of the more interesting redbud cultivars:

**'Rising Sun'** has attractive tan bark, making it distinctive from other redbuds.

**'Royal White'** has large flowers that are more hardy in cold temperatures than other "white" redbuds.

**'Silver Cloud'** has variegated foliage that looks like someone threw white paint on the leaves. The pattern varies from year to year. This variety was introduced by the late Kentucky plantsman Theodore Klein of Crestwood.

**'Flame'** is a double-flowered redbud that produces blooms on zigzag branches. It is reportedly less cold-hardy than the species.

**'Covey'** is a weeping redbud, as is the dwarf form 'Lavender Twist.'

**'Don Egolf'** is a Chinese redbud that is a very drought-tolerant 9-foot shrub. It will flower in partial shade.

Your local garden center is the best source for redbuds, but if you are looking for a particular cultivar, you may need to turn to an Internet search.

Redbuds are versatile in the landscape. Because they are small trees, redbuds can fit in nearly anywhere. Many cities and homeowners are using them to plant beneath power lines where large trees don't belong. Redbuds in bloom look fantastic planted in front of evergreens, such as a row of white pines. A dwarf form, 'Lavender Twist,' can be used for a patio or even a container plant. The Asian redbud, Cercis chinensis, is a vase-shaped shrub that literally covers itself in pinkish-purple blooms in spring and will fit perfectly in a shrub border.

Redbuds will grow in the shade but flower best in full sun—think woodland edge. And when choosing a site for the tree, keep in mind that redbuds are relatively short-lived. It is one tree you may outlive and have to remove at some point.

## Cultivation and Care

A tree that can grow out of nearly solid rock along a cut in the interstate is obviously one tough customer. But that also tells the gardener that the redbud is going to want well-drained soil. In fact, about the only way to kill a redbud without a chainsaw is to plant it in soil that stays soggy for a half-day or more. It will grow more rapidly in fertile soil, but the redbud otherwise is not demanding of soil type, fertility or pH.

A newly planted redbud will appreciate even moisture levels; water it weekly during dry spells. Once established, the redbud is easily able to withstand dry conditions. The redbud will start to bloom when it is 4-5 years old.

If you look on the Internet, you can find a list of insects and diseases that occasionally will hit redbuds, but, in fact, the trees are rarely bothered by any pests. A disease called dieback occasionally will hit a stand of redbuds, but that happens more often where the trees are growing in close proximity in the wild. Trees planted in home landscapes rarely get hit with dieback, canker or other problems. You can prune redbuds if you want to shape them, but they really need only the occasional removal of dead or damaged branches.

I told you they were easy!

## CALL THE CABLE GUY

As they mature, redbuds frequently have large limbs that grow at angles making them susceptible to splitting away from the trunk. It is one of the reasons redbuds are typically short-lived in the wild. If you have a particularly valuable, aging specimen, you can call an arborist about the possibility of cabling the limbs to keep them from splitting off. That works, too, for larger, older trees. But that is a job for a pro; don't be doing it yourself.

# Too busy to garden? Try carefree plants

Every Kentuckian is a gardener in April and May. Who wouldn't want to spend time outdoors when the sun is cozy warm and not yet hot, blooms are bursting forth, plants are fresh and vibrant green, and the soil smells, well, earthy.

Alas, May is followed by June, July and August. And heat and humidity sap our will to keep up appearances in the garden. Remember the summer's heat, with something like 30-plus days with temperatures of 90 or above. And then there are the weeds, the bugs and the stink of the pesticides we use to rid our gardens of those. It's enough to make even an avid gardener hang up her trowel.

Don't.

Instead, when plant shopping this year, look around for those specimens that can take the heat, humidity and pests—ones that don't need babysitting in June, July and August, and still reward us with blooms and lush foliage. While no plants are completely carefree, there are plenty of selections that can thrive with the little care we may be able to give them in the (relative) cool of the early morning and late evening. Plants are out there that not only require a little less of our time—sometimes a lot less of our time—but they also, in most cases, require less water, fertilizer and pesticides. So they not only make life easier, but they also make it greener. And we can admire them from the cool of our air-conditioned homes.

Below is a list of plants that we'll call "carefree." I will describe one or two plants from each group, and then list a few other species you might want to consider as you plant shop on the first warm weekend in spring. Remember, there are other hardy plants available, and many progressive garden centers mark those plants as particularly well adapted to the growing conditions in Kentucky gardens. Make friends with the garden center employees, and they will help you locate those tough plants.

## TREES

The ginkgo tree, also called the maidenhair tree, survived the age of brontosaurus and tyrannosaurus rex, and it can certainly survive the heat and humidity of a Kentucky summer. Ginkgo trees are not bothered by heat, insects or disease and tolerate pollution and a wide range of soil types. Trees typically reach 50-60 feet in height and need little or no pruning. Fall color of the uniquely fan-shaped leaves is a vibrant, buttery yellow. Try to find males of the species, though, because the female produces a fruit appreciated by gourmands of Chinese cuisine, but unappreciated by the nostrils.

Other tough trees include the golden rain tree, tulip poplar, Kentucky coffee tree, bald cypress and redbud. Though all of these are drought resistant, remember that any new tree will need watering its first summer during periods of drought.

## Shrubs

Kentucky gardeners are starting to discover the native beautyberry, ('Callicarpa') as evidenced by its increasing presence in garden centers. Ten years ago, you couldn't find beautyberry except in specialty catalogs. Beautyberry is a tough shrub that produces uneventful spring blooms followed by clusters of pink, purple and sometimes white berries that put on a show until well into winter. Yellow fall foliage is a bonus. Beautyberry looks great massed in a shrub border or as a specimen in a planting of perennials. Disease-free and drought-resistant, beautyberry may occasionally freeze in sub-zero winters, but will come back from the ground.

Other carefree shrubs are viburnum, flowering quince, rose of Sharon, sumac and fuzzy deutzia.

## Vines

Vines add stature to the garden and are beautiful in flower and form while serving as a backdrop for shorter annuals and perennials. But vines are still underused by most gardeners, probably because so many "climbers," including some roses and hybrid clematis, are finicky and drop dead at the first sign of stress. Try the native sweet autumn clematis. Its flowers are star-shaped, white and more modest than those of hybrid clematis. But what sweet autumn clematis lacks in flower size it makes up in quantity; the fast-growing vine literally covers itself in white blooms in very late summer and early fall. And the smell is heavenly. The only fault with sweet autumn clematis is that it can self-sow and become a pest. Plant it where it can't spread.

Other hardy vines are morning glory, silver lace vine, crossvine, trumpet creeper and trumpet honeysuckle. The latter two are also magnets for hummingbirds.

## Perennials

Purple coneflower (Echinacea) and

coneflower ('Rudbeckia') are native plants of the prairie and the eastern United States. Both are tough summer bloomers that shrug off heat, insects and disease. Purple coneflowers have become so popular, breeders have produced them in colors ranging from white to green to pink and lipstick red. Rudbeckia flowers are usually yellow, orange or white petals arranged daisy-like around a cone. Birds, especially finches, love the seeds of both species and will snack on the cones all winter. That can be a problem because both plants will self-sow with the help of the birds and spread where they're not wanted. Then we call them weeds.

Other tough perennials include: blackberry lily, daylily, goldenrod, Russian sage and stonecrop.

### Annuals

Cosmos is an annual that ought to get more attention. Cosmos has bright, divided, fern-like foliage that breaks up the texture of more sturdy-leaved plants. Cosmos looks delicate in the garden, but it is one tough plant. Cosmos' flowers come in colors from white to almost black, and it blooms from early summer into fall. Plant cosmos among perennials and annuals that will help it "stand up" after heavy rains, which can make the flowers sag to the ground. Cosmos does occasionally self-sow, but it isn't as invasive as the coneflowers.

Other easy-care annuals are marigold, zinnia, impatiens, Mexican sunflower and nicotiana (flowering tobacco).

Go carefree and take it easy out there this summer.

## A CAREFREE LANDSCAPE

In downtown Shelbyville, landscape designer Betsy Smith created a low-maintenance garden that borders a parking lot, the history museum and First Presbyterian Church. She used grasses, oakleaf hydrangea, a few small trees and Knockout roses. The result is a garden that is colorful, comfortable to sit in and low-maintenance. That combination of plants should work for you, too.

# Honeymoon time in the garden

Gardeners in Kentucky enter what I call the "honeymoon phase" of gardening in the spring.

Fresh from looking over the season's catalogs with pictures of perfect gardens, flawless flowers and weed-free lawns and vegetable patches, we see the gardening season ahead as one long, blissful encounter with the wonders of growing things and the beauty of nature. It is springtime, after all.

We don't see the heat, humidity, the bugs and weeds that have many gardeners longing for an apartment in the city by mid-July.

Realistically, there is no way to plan a perfect garden, just as there is no way to plan a perfect marriage.

But there are a few things gardeners can do to make the season ahead fulfilling and bountiful, if not blissful.

Let's start with the lawn. I am not a fan of spending time and money on weed killers, artificial fertilizers and insecticides. If you need a perfect lawn, hire a lawn care company.

If you plan to take care of the lawn yourself, the single best thing you can do to make your lawn look better is to cut the grass high and often. Set the lawn mower so it cuts the grass at least three inches high. A tall stand of grass will shade out most weeds, retain moisture and ward off diseases better than a scalped lawn.

Also, if you have a lawn that is primarily bluegrass or ryegrass, consider switching to one of the newer fescues, like Rebel, that is better suited to Kentucky's growing conditions. But the best time to re-seed a lawn is August or September.

If you have or are planning a perennial garden, the best thing you can do for your plants is to put them in raised beds.

Most gardeners in Kentucky have to deal with sticky clay soil that cramps perennial root growth and causes poor drainage, the bane of our most popular perennials, including roses, daylilies, peonies and hostas.

Dig or rototill the soil in the bed to a depth of 6 to 8 inches, and then add 4 inches or more of purchased topsoil or compost on top of that before putting in the plants. Then surround the bed with rocks, timbers or whatever suits you. Some gardeners just mulch the raised bed to create a berm, and that works, too.

In the perennial garden, I believe in survival of the fittest. No matter how much you want a plant, if it's one that constantly needs spraying, fertilizing with exotic soil amendments and otherwise fussing over, I get rid of it. Better to give up a plant than give up gardening.

Annuals require less extensive soil preparation, but they are particular about where they are put in the landscape. Some tolerate full sun and blistering heat; others want cool, damp and shady.

Annuals that can take the heat include marigolds, petunias, salvia, geraniums, and begonias, though some of the begonias with dark leaves fade in the hot afternoon sun. The best annuals for shade are impatiens and vincas, both easy-to-care-for specimens.

If you are planning a vegetable garden, the best advice I can give you is be realistic about how much you can take care of come July and August. I always give out this advice, though I have never taken it myself.

Remember, if you plant the entire seed packet of zucchini you will be harvesting the green monsters by the truckload by July.

Remember that a well-grown tomato plant can produce as much as a bushel of tomatoes; keep that in mind before you leave the garden center with five six-packs of plants.

Finally, if you want to garden by the book, a really excellent resource is the *Southern Living Garden Book*, edited by Stephen Bender, one of the best garden writers in the country.

So much information about gardening in this country is geared to someone who lives on a latitude of about Dayton, Ohio, or Indianapolis. This book has everything you need to know and is geared to the gardener who fights the heat, humidity and diseases endemic to the South.

And last, remember that, like marriages, no two gardens are alike. Read all the advice you can get, and try some of it. But do whatever makes you happy.

## SUMMER SHADE

Many gardeners make the mistake of thinking that if a plant requires full sun it can't stand any shade whatsoever. The truth is, most plants will do quite well with just some morning and early afternoon sun and appreciate a respite from the scorching rays of the sun in mid-afternoon, especially here in Kentucky. That's why it's easier to find plants for an east side of the house exposure rather than a west side. If you want a west-side garden, try placing some lacy-leaved large trees away from the garden to provide some dappled shade. Trees such as river birch, locust and willow work well.

# Five fabulous perennials

Any gardener worth her trowel likes to try a few finicky plants: the hybrid tea rose that needs constant spraying, the azalea that is persnickety about soil type, or the delphinium that demands near-perfect weather. But the mainstay plants in my garden are those that are tough and need little care.

Let me introduce you to five perennials that fit the "tough as nails" category. They grow in most soils, thrive without constant attention, and look good in bloom and not in bloom. I suggest you give one of these fabulous five a try and you will become a fan too:

## Columbine (Aquilegia)

Columbine flowers look like pansies in flight. Spurs behind the flower give an appearance of being ready to "lift off." In fact, the Latin name means flight of eagles and the common name, columbine, comes from the Latin for dove. The flowers come in a range of colors. The McKana hybrid varieties will be the most readily available and also contain the widest assortment of colors.

The columbine plant ranges from 18 to 30 inches in height and looks like an overgrown clover. After the flowers are gone, the plant remains handsome throughout the summer. Columbine will survive in nearly any soil, but a humus-rich soil will bring out their best. Columbine will tolerate full sun but prefer light shade.

## Coreopsis

Also called tickseed, this member of the daisy family will come as close as any plant to blooming all summer, demanding only an occasional deadheading to remove spent flowers. Most coreopsis flowers resemble daisies, but the plant remains shorter than daisies, 8 to 24 inches, and doesn't get as scraggly.

I planted 'Early Sunrise' coreopsis several years ago and had more requests for plants from that patch than from anywhere else in my garden. Their bold yellow blooms coming in the first of June shout, "It's summer!" Another favorite coreopsis is 'Moonbeam,' a light yellow. A similar variety, 'American Dream,' is pink. Plant coreopsis in ordinary soil and full sun.

## Mallow (Malva)

Mallow is in the hollyhock family, but it has all of the hollyhock's virtues—striking flowers arranged on a tall stem—and none of its drawbacks. Mallow is not liked by Japanese beetles and is not bothered by rust that turns hollyhock leaves brown.

Mallow is technically a biennial, meaning it makes a plant one year and blooms the next. But because mallow self-sows so easily you will have blooming plants every year. In fact, if there is a drawback to this plant, it is that it can take over a garden easily.

Mallow's large, flat leaves remain a deep green all summer. The trumpet-shaped flowers are usually red, yellow or pink. Plant in full sun.

## Spiderwort (Tradescantia)

Don't let the common name of this plant put you off. Spiderwort is not particularly loved by spiders, though your guests who see it enhancing a shady spot in your garden will find it attractive. Spiderwort is a carefree plant that loves the shade. In fact, its flowers will close up when the sun strikes.

Spiderwort is a grassy plant with blue-green arching leaves that grows about 2 feet tall. In early to mid-summer, dainty flowers, usually in shades of blue, will appear and persist most of the summer. When the flowers are gone, the rich, blue-green leaves carry on the show. Spiderwort is easy to transplant.

## Yarrow (Achillea)

Yarrow is as prized for its silver-gray, ferny foliage as for its flat, dainty flowerheads that appear to float above the plants.

Yarrow thrives in hot, sunny spots and blooms most of the summer. Wild yarrow blooms white but most cultivated yarrows are yellow, the most satisfactory of which is 'Coronation Gold,' a cultivar that grows to about 3 feet.

Because of their height and foliage, which contrasts with most garden plants, yarrows look great planted among lower blooming plants. Try putting yellow yarrows behind or among purple salvias then sit back and watch a show that will last all summer.

## NO THANKS

Many of the most popular perennials, including daylilies, hosta, yarrow, peony and iris, can be propagated by digging up pieces of existing plantings. Do not be afraid to slice off pieces for sharing with other gardeners. The perennials actually will perform better if they're thinned occasionally. It's your garden's way of telling you to be generous and share. Just remember that if you're the recipient of a gift from the garden, it's bad luck to say "thank you."

# Hail, Iris, goddess of the rainbow

A springtime trip that once took me down Ky. 79 toward Morgantown also became a trip down memory lane.

In many of the yards along the way I was thrilled to see purple iris, what we used to call "flags," blooming in a circle, probably planted many years ago when tire circles were all the rage.

When I was a kid nearly everyone had a circle of "flags," usually out front, sometimes in front of a row of 'Festiva Maxima' peonies. The flags bloomed in mid-spring, always in some shade of purple, from lavender to deep violet. The plants were tough and the blooms dependable year after year with little care and attention.

Today, hybridizers have produced iris with huge flowers in nearly every color of the rainbow except true red. Iris got its name, in fact, from the Greek goddess of the rainbow who, in Homer's *Iliad*, delivered messages using the rainbow as a means of travel between heaven and earth. The iris' nickname "flags" comes from the flower's use on the French flag as a symbol of the Bourbon royalty.

The iris can be roughly divided into two types: bearded and non-bearded.

The bearded types, or German bearded iris, are the "flags" of my childhood and probably the plant most people think of when they think of iris. The plant has a wide blade, usually a sea-green color. But striking variegated (green and white) varieties are also available.

German bearded iris flowers have upright petals, called standards, and drooping petals called falls. The standards and falls can be the same color or different. On each fall is a fuzzy "beard" that resembles a caterpillar in size and shape. German bearded iris come in sizes from dwarf, no more than a few inches tall, to large, 3 feet tall or more.

The German bearded iris displays the largest range of color in the iris family, probably because it has received the most attention from hybridizers. 'Christmas' is an outstanding white. 'Greenman's Castle' is a prolific blooming peach color. 'Role Model' is a vibrant lavender that is stunning when backlit by the sun. 'Study in Black' is such a deep, royal purple it is nearly black. Combine 'Study in Black' with 'Christmas' or place it in front of white peonies for a show-stopping combination in your garden.

The nonbearded iris include the Siberian, the Japanese, the species iris and the "flag" iris. For the most part, their flowers are smaller and more horizontal than bearded iris and they come in a narrower range of colors. Siberian iris, for example, appears in blues, purples, whites and yellows. 'Caesar's Brother' is a popular purple Siberian. 'Gull Wings' is a popular white.

Many people prefer non-bearded iris to bearded because the former's foliage is more grass-like and easier to fit into the landscape. Another advantage of many of the non-bearded is that they will grow in wet spots; some, like the yellow flag and blue flag iris can grow in several inches of water. Some people plant them in their water gardens. Non-bearded iris also tend to be nearly disease- and insect-free, while bearded iris can be plagued by rots and iris borers in some climates.

## Care

Summer is the ideal time to plant bearded iris. Buy from a catalog, an iris breeder or, better yet, find a friend who has a clump that needs dividing. Help with that chore and take a few rhizomes home with you for your efforts.

Bearded iris has to be the easiest plant

in the world to establish. I have thrown rhizomes out on the garden, even on the lawn, and had them take root and start to grow. But the better way would be to prepare a bed by working the soil at least several inches deep and placing the rhizomes in the bed 4 to 6 inches apart. Don't bother to bury the rhizomes; in fact, iris breeders say they should sit on the ground "like ducks on a pond." The leaves may shrivel after planting, but after a couple of months, new growth will take off. About half the irises you plant one year will bloom the next. The plot will be in full bloom by year two.

Keep the beds weeded. If the soil is reasonably rich, lay off the fertilizer. Watch for signs of borer damage, usually indicated by rotting rhizomes shot full of small holes. Dig those out and destroy those plants. You might want to dust the remaining rhizomes with an insecticide.

Non-bearded iris can be set in spring in holes as you might other perennials. You might want to sprinkle a handful of fertilizer around them after they get growing but don't overdo it. Make sure non-bearded iris plants stay well watered; in fact, don't be afraid to put them in wet or swampy places.

Whether bearded or non-bearded, iris eventually need dividing to keep flowering at their best, usually after three to five years. It's usually best to dig up the whole bed, break apart the rhizomes or plants and start over with plants 4 to 6 inches apart.

And don't try throwing the rhizomes out on the ground to get rid of them. They'll just take root, grow and flag down another generation of gardeners.

## FLAGS

When we were kids, we called all irises "flags." I have no idea why. But there really are two irises that have "flag" in their name, and they are good choices if you want a native plant and one that would do well in a wet spot. They are the blue flag and yellow flag iris. Both thrive in moist soil, even standing water. Yellow flag is larger and a bit more robust. Unlike the German bearded irises we called flags, the yellow and blue flag iris have a more orchid-like flower. Try them. They're carefree and hardy.

# Suit yourself

My garden will never appear on a list of garden tours.

My gardens have always featured an eclectic mix of plants that would scandalize any self-respecting garden designer.

I have roses cohabiting with German iris, daylilies, nandina and various and sundry bulbs.

In the front border, I have some respectable yews, boxwoods, birches and a 'Jane' magnolia; all have the Good Housekeeping seal of garden design approval. But I also have hostas, iris, daylilies, mums and whatever summer annuals my wife can't resist at the two-for-one flat sales.

Part of the problem is that I get lots of free-plant offers as a member of the Garden Writers Association. So when they arrive this spring I will plant the All-America 2003 winning roses somewhere. I'll find some place to plant the free gladiolus bulbs that arrive around Derby Day. And come fall, I'll figure out where to tuck in the bulbs I buy with my Dutch Gardens gift certificate.

But the main reason my garden won't make the tour is that long ago I weaned myself from an overemphasis on garden design.

About 10 years ago I went through a phase where I read every book I could get my hands on about garden layout and design. I've got a shelf full of books that discuss nothing but the use of color in the garden. Books on the use of plant height and depth. Or the effective use of foliage patterns.

And I learned most of the rules. Like tall in back, short in front. Put yellow with purple. Use white to separate colors. Blue and white gardens are "cool"—the temperature, not the quality of hipness. Red and yellow gardens are "hot"—the temperature, not the sexual attractiveness.

And on and on.

Once, inspired by pictures of the "white" gardens of England's Sissinghurst Castle designed by Gertrude Jekyll, I decided I wanted a black and white garden.

I searched high and low for plants with blooms as pure white as possible, which are not as easy to find as you might think. I looked even harder for black plants, or at least plants that were so red or purple they appeared to be black. And I mixed them together in a 15-by-4-foot bed.

It didn't work.

The purples and reds clashed violently. The whites faded and refused to cooperate by blooming at the same time as the reds and purples. And for much of the season, the garden was little more than a patch of green, with nothing blooming at all. In my zeal to achieve a color pattern, I got plants that finished their blooming by mid-June.

These days, partly because I stay very busy, I put plants wherever I can find a space. I do pay attention to the plant's ultimate size and growth habit and its need for sun or shade. I don't, for example, put a maple tree next to the driveway or a rose bush on the north side of the house where it will go begging for sun. But otherwise, I let the colors riot where they may.

Sometimes the result is an occasional color faux pas. The pale red zinnias just made the pale yellow marigolds nearby look kind of sickly. The bright pink roses next to bright yellow ones get along about as well as siblings on a long road trip.

But I also got some happy surprises, like the picture painted by the bed of grape hyacinths blooming below clusters of thalia daffodils, their heads nodding down at the hyacinths as if saying to their purple companions, "Don't we look spiffy together!"

When I get asked about designing gardens now, I suggest planning two, no more than three, plant combinations.

For spring, try yellow or white tulips planted among a bed of grape hyacinths. If you want another element and have the space, plant a flowering magnolia in the middle of the tulips and hyacinths.

For summer, try yellow marigolds or mixed zinnias fronting white coneflowers. Put them in the sun and they will last until frost.

For fall, try yellow or purple mums planted in front of white or yellow roses.

Once I gave Mrs. Sweeney, a former neighbor and one of my favorite people, some of my German iris rhizomes. When I went over to her house a few days later, she was standing in her flower bed, holding her hoe and fretting over whether to put the purple ones next to the white ones. Or should she put them next to the yellow ones?

"Mrs. Sweeney," I said. "When a beautiful person like you plants a garden, no matter what you do, it will be beautiful."

## SOMETHING TO AGREE ON?

Even though it's my position that you don't tell adult gardeners what two colors can or cannot go together, studies show that combinations of certain colors are almost universally distasteful. For some reason, human eyes respond negatively to clashing pink and orange shades yet like the combination of golden yellow and orange. Now you know!

# 10 tips for your best-ever vegetable garden

The vegetable patch can be the best of what we gardeners do. Or it can be the worst.

At its best, the vegetable garden produces abundantly and looks attractive at the same time. At its worst, the vegetable garden becomes a weed-infested mess that nets little in return for time and labor invested.

When the growing season gets under way in earnest, try these tips for making your vegetable garden as productive, and good-looking, as it should be.

## 1. Choose the right seeds and plants

Vegetable varieties vary greatly in productivity, insect and disease resistance and adaptability to our climate. Some varieties will do well in western Kentucky but perform poorly in the mountains. Unfortunately, many seeds sold on the racks in stores and garden centers are not of varieties that perform well here.

Notice what varieties you plant and stick to the winners. Experiment with new varieties alongside the tried and true. Visit your Extension agent for a list of varieties that should do well in your area.

## 2. Choose the right site

Planting vegetables near trees is an exercise in futility for two reasons. Vegetables need at least six hours of full sun daily, and trees will throw too much shade. And trees will out-compete vegetables for water.

Planting in a low spot, where water stands, is no better. The ideal spot for a vegetable garden is in full sun or in a spot where shade comes only in the afternoon. The ground should slope slightly to provide drainage.

## 3. Watch your timing

Midspring is probably already too late to plant cool-season vegetables such as broccoli, cabbage and lettuce. Best to wait until late August or early September to catch the cooler weather of fall.

Warm-season vegetables, such as tomatoes, melons, cucumbers and squash, can be planted in most of the state early in May. But watch the weather. If a strong cold front is on the way, a day or two of frost may follow. Let that pass and by mid-May, it should be safe to plant the warm-season crops.

### 4. Reach for the sky

Wherever possible, grow up, not out. Plant varieties of beans that are called "pole," meaning they grow 6 feet tall or higher. Plant them on the traditional stick tepees or use bamboo and string to create a lyre effect. Plant tomatoes, even cucumbers and melons, in cages or along fences and train them to grow up. The plants stay disease-free, the fruits are cleaner and picking is easier. Melons would need some kind of support; use old pantyhose.

### 5. Go easy on the fertilizer

Imagine eating one huge meal in May and nothing again. That's what we ask our plants to do when we put down a gob of fertilizer at planting time and nothing else as the plant produces. Here's a general rule: Fertilize at half the rate recommended on the bag when the seeds or plants go in the ground. Add the other half when fruits start to form. Fertilize lightly again after the plants have been harvested if you want them to keep producing. Another approach is to fertilize with a natural, slow-release fertilizer, such as compost, which will break down slowly and feed the plant over the entire growing season.

### 6. Keep up with weeds

Use mulch once plants are up and growing or hoe diligently when weeds are small. Once weeds get large, the good plants are often damaged when the weeds are taken out. And never let weeds go to seed. A single pigweed, a common garden nemesis, is capable of producing as many as 1 million seeds, seeds that will haunt you for the rest of your gardening career.

### 7. Troll for bugs

The Chinese say the best fertilizer is the farmer's (or gardener's) shadow. Walk through the garden frequently, checking leaves, especially the undersides, for insects and insect eggs. Take a jar of soapy water with you and let the little guys get good and clean before they drown. If you are too squeamish for that, use insecticides before the bugs do their worst damage. Vine crops—melons and cucumbers in particular—can go from green to gone within days because of cucumber beetle damage. Know thy enemy and zap him before he gets a toehold.

### 8. Use flower power

Who says a vegetable garden has to be just vegetables? Flowers can dress up a row of beans or broccoli, and they do double duty by deterring some insect pests. Marigolds, nasturtium, zinnias, sunflowers, hollyhocks and salvia are good garden companions.

### 9. Plant in succession

As plants produce heavily, they gradually succumb to insect damage and disease and just plain wear out. Keep fresh plants coming up to replace the old-timers. In other words, if you have a packet of zucchini seed, do not plant all 25 or 50 seeds at once, unless you are going for commercial crop status. Plant two or three hills of zucchini in early May; another two in early June; three more in July, and a couple more in early August. This works for cucumbers, corn, beans and tomatoes.

### 10. Keep notes and records

Note what varieties produced well, tasted the best, resisted disease, grew best during dry spells. That will come in handy when you order seeds next year. Do you really think you will remember the summer's garden in January? Also, draw a sketch of the garden and what was planted where. Ideally, vegetable crops should be moved around every year to avoid disease buildup. Your sketch will be your map for planting next year.

---

## TWO ARE BETTER THAN ONE

Garden sprayers are relatively inexpensive, and I recommend keeping two on hand, rather than one. One should be labeled for herbicides, the other for insecticides and fungicides. If you don't thoroughly clean your sprayer after using an herbicide, the next time you spray a plant for insects, you may kill more than the bugs.

# Made in the shade

I wish I had a sack of fertilizer for every time I hear a gardener whine about having "too much shade in my backyard to grow anything."

I, who garden almost exclusively in all-day sun, think of shade as a blessing rather than a curse. True, the number of plants that grow well in shade is more limited compared with those that want to bask in the sun, but limits are a hedge only to the uncreative or lazy mind. Some of the most beautiful and serene gardens I have seen, in real life and in pictures, have been shade gardens.

Think of shade as an ally rather than an enemy. Where would you rather sit on an August evening, in a shady garden or in full sun? Shade keeps the colors of flowers more vibrant because the sun does not fade them. Colors such as pink, lavender, yellow and white, show better in the shade than in the harsh light of summer sun. In the shade, watering may be infrequent or unnecessary except during the driest spells.

Dealing with shade effectively requires first knowing how much shade you have and where it comes from. Deep shade from buildings or trees requires a different planting strategy than dappled shade; many plants that call for "full sun" in the catalogs actually tolerate dappled shade very well in our climate. Shade from the west side of a building is a different critter than shade from the east side. And remember that what is shaded changes as the season progresses; those deciduous trees that cast deep shade in summer are bare in the spring. Go ahead and plant spring bulbs, such as daffodils, Virginia bluebells and early tulips. They will finish their flowering before their world is plunged into the darkness of a large maple.

Shade from buildings is less of a challenge than shade from large trees, not because the quality of light may be different but because large trees suck up fertilizers and moisture and leave little for shrubs and perennials below without your intervention. In the battle for nutrients and water, big trees win every time. Maples are especially notorious for casting almost impenetrable shade and creating a scorched earth beneath their feet. If your yard is full of sugar or Norway maples, the best strategy may be to plant a tough groundcover, such as pachysandra or ivy, at their feet and be done with it. I know a gardener who plants stinging nettle beneath the shade of a water maple because that is one plant that will grow amid the roots that rise from the ground like corpses in a horror movie.

If you are not dealing with large trees, the number of options becomes infinitely larger. Remember that when planting a shade garden, texture of foliage may be as important as flowering. The round leaves of most hostas contrast nicely with the arrow-shaped leaves of ferns, for example. Here are a few ideas for shady gardens you may be planting this spring:

Suppose you want a small, perennial garden on the north side of the house where a neighbor's house also blocks the sun. You need plants that tolerate full shade. Try a cinnamon fern or a large blue hosta such as 'Bressingham Blue' as the centerpiece. A goatsbeard (Aruncus) can go in the back next to the house: it will bloom white in mid-spring. Around the front and sides arrange black blooming sedge grass (Carex nigra) or variegated Japanese sedge (Carex hachijoensis 'Evergold') for all-summer interest.

Now let's try a small garden for the east side of the house where trees block some, but not all, of the morning sun. Shade will be full by mid-afternoon. Pick plants that tolerate either full or partial shade.

If the garden is sizeable, plant a rhododendron next to the house if drainage is good. If not, plant a clethra, a shrub that will tolerate considerable shade and still bloom well in summer with flowers that will perfume the area. 'Hummingbird' is a good variety. Plant two or three large hostas such as 'Frances Williams' or 'Francee' at the shrub's base. You can plant impatiens in front of the hostas. Foamflower (tiarella) is a low-growing perennial that will appreciate the cool shade.

If the part-shade garden is smaller, try three monkshood ('Aconitum') plants in the back, next to the house. If your house is white, go with the blue or purple monkshood; if dark, try the white varieties. In front of the monkshood, try some Japanese painted ferns and a low-growing variety of astilbe, such as 'Pumila.'

One more. Suppose you have a garden on the west side of the house. You get morning shade but a full blast of afternoon sun. This may be the most challenging garden because you get sun when you least want it and shade when you least need it.

For this garden, try a mixture of the yellow or gold-leaved hostas, which can take a shot of sun and heat better than their blue and green cousins. Plant 'August Moon,' 'Sum and Substance' and 'Piedmont Gold.' Mix in a little of the carex sedge grass for a different color and texture and add some toad lilies for their late summer, orchid-like blooms.

Put in a bench for sitting and you will have it made in the shade.

## SHADE SHRUBS

It is relatively easy to find perennials and annuals that enjoy shade; it is not so easy to find shade-loving shrubs. But they are out there, and there are some good ones: hydrangea (macrophylla, quercifolia and arborescens) fothergilla, ilex, itea, leucothea, pieris, mahonia and viburnum. While these shrubs won't appreciate deep shade, they will perform quite well with just a few hours of sun or a half-day of dappled, indirect sunlight.

# First fruits: strawberries

One of my earliest memories is of visiting my cousin Cliff Yadon's farm in Bowling Green late in May, about strawberry picking time.

I was too young to be invited to pick in the patch, but I was on hand for the grading and packing. The berries would roll down a metal conveyor and many hands (not mine) would pluck them up and put them into the wooden quarts. The quarts went into wooden flats and the wooden flats set aside when full to await shipping, or a little boy's fingers filching one or two.

Strawberries are the first fruits of the garden, coming in mid- to late May when raspberries, melons and peaches are a still-distant dream. I have never grown strawberries on the scale Cliff did, but I have always planted a patch. If you have never grown your own strawberries, you do not truly know what a strawberry tastes or smells like. If you do not grow your own strawberries and make your own jam, you do not know what strawberry jam can be without all of the cornstarch and preservatives.

Strawberry planting time in Kentucky is in April. Plants are available in garden centers, and it is probably not too late to order them from a catalog or online sources. Catalogs will offer dozens of varieties of strawberries. June bearers are the most common strawberry. They will yield a large crop in late spring and be finished until next spring.

'Early Glow,' 'Redchief,' 'Guardian,'

'Delmarvel,' and 'Surecrop' are popular June bearers in Kentucky gardens.

Ever-bearing strawberries and the so-called day-neutral strawberries produce berries in small numbers on and off during the summer. 'Tristar,' 'Ozark Beauty' and 'Fort Laramie' are ever-bearing and day-neutral strawberries. I would recommend the June bearers over the ever-bearers because unless you have a very large patch, you never get enough ever-bearers to make more than a handful, much less enough for a strawberry shortcake.

One thing you should know before planting strawberries: You are going to need some patience. The strawberries you plant in April will not produce until next May. Yes, the plants will produce a few strawberries the first year if you fail to clip off the buds, but letting them bear the first year will badly cut your yields next year.

Good strawberries start with good ground. Strawberries like a loose soil that is high in organic matter. A 3-foot by 25-foot row will accommodate 25 strawberry plants, more than enough for a family of four. Till or spade in some manure from a horse farmer, compost or purchased manure from a store.

Your strawberry plants will come wrapped in a rubber band and may look to you like nothing more than a mass of hair, something like horticulture's version of ZZ Top. If you pull the plants apart and look closely, you will see the mass of roots extends out from the plant's crown, a light green swollen nodule that sometimes has a leaf or two attached.

Now quit playing with the plants and put them in a can of water until you are ready to plant, but do not keep them in water for more than 24 hours. With a trowel, dig a hole in the soil, fan out the roots with your hand and fill in with dirt. The crown of the plant should rest just at soil level. Give the plant a squirt of water and go on to the next plant, which you will place about a foot away.

During the first year, you want your plants to produce runners, not berries. So remove all blossoms as soon as they appear. When the runners start appearing, you can maneuver them around the mother plant in a fan or star shape. Allow four to six runners per plant and they should fill in the row satisfactorily.

Besides not getting berries the first year, the most frustrating aspect of raising strawberries is keeping out the weeds. Weeds are probably the No. 1 reason people give up on strawberries. As the runners fill in between the plants, hand-weeding becomes tedious, almost a matter of pulling weeds out individually.

A couple of strategies will help combat weeds. One is the use of raised beds with sterilized soil. Raised beds will give the strawberries the drainage they need, and weeds will not show up in the sterilized soil, at least not until later in the summer. Another strategy is to try a granular herbicide, such as Preen, at planting. If the directions are followed carefully, the herbicide will keep most weeds out of the patch, but expect to need a second application by mid-to late summer.

In the fall, it's a good idea to fertilize the plants with a 5-10-5 formula to encourage strong root and fruit formation the following spring. The "straw" in strawberries comes in late fall after the first or second good freeze. Shake a bale or two of straw loosely over the plants in late November or early December. You want to cover the plants but not smother them. The plants should still be barely visible beneath the straw before their winter nap.

In the spring, pull back the straw when the danger of hard frost is over, usually in late April. But keep the straw handy in case of a late freeze. The straw will also keep the berries cleaner. Birds are attracted to the color red and may beat you to the berries just when all of your patience is starting to pay off. Try covering the row with floating row covers available at garden centers.

One problem with home-grown strawberries is that, unlike their shipped-in, cardboard cousins, they will not keep very long. You have to eat them, make jam or freeze them within a day or two. But that's a good problem to have.

## WEED FREE

The biggest pain in growing strawberries is keeping out weeds. One cure is the so-called "strawberry pot" found in most garden centers. The pot is filled with soil, and the plants are put in openings along the side. The strawberries grow clean and weed-free. The pot will have to be kept in a garage or enclosed building to overwinter, however.

# Marry plants for a razzle-dazzle garden

For the last 10 years or so, a trend in landscape design has been to plant large drifts of all the same plant, all the same color. Such a planting can create some stunning vistas, true, and banks of flowers help lead the eye from one part of the garden to another. But when the goal is to make the garden visitor stop and say "Wow!" try pairing plants in eye-catching, compatible marriages of color and form.

The easiest way to create a compatible pair is to combine two colors of the same plant, a yellow daylily and purple daylily, for example. Or red and white marigolds. But the mating of two different plants of two different colors allows the gardener to work with color, shape and size to create some matches made in heaven.

A few suggestions: First, make sure you select plants that bloom at about the same time and are happy in the same space. Red tulips and yellow marigolds could be a pleasing combination, for example, but alas, the tulips will be long gone before the marigolds even emerge from the ground. A May and December marriage!

And join in matrimony plants that will be happy together. Don't place a shade lover with a sun lover, or a water-loving plant with one that likes it on the dry side. Pink yarrow and blue monkshood, for example, could be an effective combination in the garden because of their colors, but yarrow likes a hot and dry spot while monkshood likes it cool and shady. That combination would be like trying to make a tropical islander happy living with an Eskimo.

Now for colors. I've never been a color scold, but it is true that some colors do look better next to one another, and if you are going to arrange a plant wedding, you ought to find colors that can get along.

Here are a few suggestions on colors that blend well from the book *The Perennial Garden: Color Harmonies through the Seasons* by Jeff and Marilyn Cox: combine deep red with light blue; combine deep purple with light yellow or light purple or lavender with deep yellow; try blue with orange; mix deep pink with light blue; and blend white or silver with anything. In fact, if you do have clashing colors, say, light purple and light red, putting white in between will keep peace in the family.

Another thing when planning a marriage—think about foliage. Large or coarse-leaved plants blend well with plants that have finer foliage.

Finally, choose both plants for hardiness. A garden marriage is rather sad when the bride or groom has met an untimely demise.

When you are at the garden center picking out spring and summer plantings, you might want to try some of the combinations below:

### Artemesia with purple salvia

Planting 'Silver Mound' artemesia between clumps of 'May Night' salvia creates a "cooling" effect in the garden in mid-summer when both plants are at their best. Plant in full sun.

### Roses with baby's breath

Most rose gardeners plant roses in beds by themselves because the foliage needs air circulation to keep down disease. But baby's breath has light, airy foliage, and its small white blossoms will blend well with any color of rose. Plant in full sun.

### Sedum and chrysanthemums

Plant 'Dragon's Blood' sedum in front of a white chrysanthemum for a late summer-early fall show. Put the two in full sun.

## Monkshood and astilbe

Most monkshood, or aconitum, cultivars are deep blue to lavender. They would combine well with either pink or white astilbe and bloom in mid-summer. Both like shade to partial shade.

## Astilbe and hydrangea

Another pair of shade lovers is the combination of astilbe and hydrangea. Plant a white or pink 'Nikko Blue' hydrangea bush. If your hydrangea blooms pink rather than blue because of more alkaline soil, no problem—just select a white or lavender astilbe as its partner.

## Shasta daisies and daylilies

Plant white Shasta daisies among yellow daylilies. The white offsets the yellow, and the grassy, arching foliage of the daylily contrasts nicely with the more upright and "stemmy" daisy. Plant in full sun and they will bloom in early to mid-summer.

## Yarrow and catmint

The sulphurous yellow of 'Moonshine' achillea, or yarrow, creates a stunning combination when in full bloom with catmint, or nepeta. Plant in full sun.

## Cosmos with ribbon grass

Not all effective marriages have to be flowers with flowers. Flowers with foliage works well, too. One beauty is the combination of 'Cosmic Orange' cosmos with the white and green variegated foliage of ribbon grass. Plant the light, airy cosmos in back of the nodding and more substantial ribbon grass. Plant in full sun or light shade

## Petunias with rudbeckia

One of the most striking combinations I saw last summer was a planter full of overflowing 'Purple Wave' petunia set amidst the black and yellow blooms of black-eyed Susans, or rudbeckia. Plant in full sun for a season-long show.

## SILVER IS GOLD

The color silver goes with about anything in the garden. No, you don't have to set out the tableware. The perennial artemisia, often sold as "wormwood," provides a wonderful dash of silver and looks good between blue, red, purple or yellow flowers. The annual called "dusty miller" is a great source of silver and looks good planted with annual red or purple sage.

# Homegrown corn: sweet!

I call homegrown sweet corn the great economic leveler. You can be a billionaire, able to buy nearly anything—or anyone—in the world. But a billionaire doesn't have enough money to buy the so-good-it's-almost-painful flavor of sweet corn fresh from the garden. Nope, the billionaire has to grow his or her own—and so do you.

Too many gardeners pass on growing their own sweet corn because they think it takes up too much space. The truth is, sweet corn will produce a surprising number of ears in even small gardens. My father-in-law used to plant a patch of sweet corn in the middle of his downtown Louisville garden and every year harvest enough to freeze. Many people who try to grow sweet corn in small spaces err in planting one or two long rows, rather than a block of several short rows. You see, corn is wind-pollinated, and when you plant a long, single row, the pollen blows everywhere except onto the tassels of another plant; no pollens on tassels means no kernels on the cob. In other words, plant four 10-foot rows of corn and you will get a substantial crop; plant one 40-foot row and you'll get almost nothing.

## Heavy feeder

Sweet corn is what horticulturists call a "heavy feeder"—it requires fertile ground to produce at its potential. If your soil is not fertile, you can feed corn with organic or commercial fertilizers to make up the difference. The good news is corn is not picky about soil type and will do well in our heavy clay as long as the soil does not stay waterlogged. Also, be sure the corn patch gets plenty of sun; corn will grow in shade, but the ears will be small or nonexistent.

Corn can be planted safely from early May through mid-June in all parts of Kentucky. You can plant corn about every two weeks to ensure a continued supply from midsummer through early fall. Alternately, you can sow early-, mid- and late-season varieties. Wait until the ground is dry, work the soil until smooth with a rake or rototiller and add organic fertilizers, such as compost or dried manure. To ensure a good stand of stalks, plant a seed about every 4 inches

SPRING 71

and then thin out the row after the plants germinate so that you have one stalk about every 8 inches apart.

Even if the ground is fairly fertile, side-dressing corn (putting fertilizer next to, but not on top of, the growing plants) is a good idea. If you use organic methods, you can use blood meal or cottonseed meal; otherwise, sprinkle 10-10-10 alongside the row when the stalks are about knee-high and again when the corn starts to tassel.

The old folks call weeding corn "laying by" and Kentucky writer Jesse Stuart had a great short story about his prowess with the hoe in the corn patch "laying by" corn. Read that for inspiration because corn does need weeding, at least until it reaches about 2 feet tall, when it will start to shade out weeds. Be careful when weeding—the roots of corn are very shallow and you can easily damage the plant by cutting too close. I like to "hill up" while I'm weeding, which is pulling dirt from mid-row up to the stalks. That offers some protection against the plant blowing down in summer storms. If your corn does blow down after a strong storm, do not despair. It will stand back up if the stalks are not laden with ears. If the ears are already on, go into the patch after it dries a little and stand the stalks back up and tamp dirt back around the roots. The plant is surprisingly resilient.

Sweet corn is ready to pick when the tassels have dried and the ear feels full all the way to the tip. If in doubt, you can pull back a couple of leaves—called the shuck—from the tip and take a look. If the kernels there are plump and full, it's ready. But don't overdo the checking because bugs frequently enter the ear where you pull back the shuck.

## Sweet, sweeter and sweetest

The sweet corn we ate as children was what my grandfather grew to feed the donkeys and cattle. We just ate the field corn—which we called horse corn—while the kernels were immature. If we had then the many wonderful varieties that are on the market today, we would have thought that we had died and gone to corn heaven. Plant breeders have produced sweet corn varieties so much sweeter than the old horse corn we used to eat. And because the varieties hold their sugars much longer, you don't have to get the pot boiling before heading to the patch—although fresh from the garden to the table is still best.

Still, I think breeders of sweet corn varieties have gone overboard in some cases. The SH and SH2 varieties, in my opinion, are way too sweet. If I want candy, I'll eat a Hershey bar. If I want a vegetable that tastes like corn, I plant the old hybrids or the SE varieties. 'Silver Queen,' a hybrid more than 50 years old now, is still hard to beat. The SE varieties, like 'Ambrosia,' 'Miracle' and 'Bodacious,' also are very sweet but still "corny," and they'll hold well in a refrigerator for several days. I like the combination of the hybrids 'Early Sunglow' for smaller but earlier ears, and 'Silver Queen' or 'Kandy Korn' for large, later ears.

Corn is a tough plant and prone to few insect and disease problems. The biggest problems are the worms that invade the ear and, while not destroying it completely, can put you off your feed when you pull back the shuck and a fat, little green wriggler flops into your lap. Earworms camp out near the tip of the ear. Corn borers are smaller worms that look, unfortunately, like a maggot and usually dig into the middle of the ear. You can spray with the organic insecticide Bt, but timing is critical—once the worm is in the ear, sprays do not penetrate. I've found that both insects are much more troublesome for late-planted corn than the early crop, so you may try early varieties if the worms destroy too many ears.

Probably the single biggest headache with sweet corn is getting it before raccoons do. The masked thieves have a knack for harvesting your corn the night before you planned to pick it for that special dinner. Several of them can destroy a whole patch in a single night. Gardeners have tried everything from electric fences to tying out the family dog to playing a radio tuned to obnoxious music to keep out the critters, and all of those strategies work—sometimes. Your best bet may be to take a lawn chair and blanket and sleep out under the stars prepared to do hand-to-hand combat with the bandits if necessary. Or better yet, pitch a tent out there for the kids and tell them they're going camping.

Whatever it takes—sweet corn is worth it.

## DISASTER!

Pride goeth before a fall! I once wrote a column on sweet corn the same year the local raccoon population took out my entire planting. Usually they would hit late enough that I would get some corn and they would get most. That year the raccoons moved in early, stayed often and wiped out my entire planting. Didn't get an ear. The next year I surrounded the patch with a cordon of electric wire. The corn might have cost $8 an ear, but the raccoons didn't get any.

# Enjoy the fruits of your labor

I always say if you don't want to do something, don't do it, but don't tell others they can't do it.

One of the most insidious rumors spread about gardening in Kentucky is that you just can't grow good tree fruits here. I've even heard so-called garden experts and horticulturists with ponderous degrees after their names say, "Oh, you can't plant (fill in the blank) peaches, apples, pears, plums here. Get them from the grocery store."

Nonsense. I'm not here to tell you growing fruit is a plant-trees-and-ignore-them operation, but neither are they the most trouble you'll ever get yourself into. (That would be orchids.) Yes, there are some tree fruits you'll have to forget about—bananas, for example. And if you want to grow apricots or sweet cherries, well, good luck. But apples, sour cherries, plums, pears and peaches can and should be grown by anyone with even so much space as a patio or balcony. Yes, there are fruit trees, mostly apples, that grow as a single leader and will easily fit onto a narrow patio or deck.

Some peach varieties are tiny little trees that will produce full-size fruit on a plant that will thrive in a 5-gallon container.

Even if you don't want the fruit—and I don't know why you wouldn't—fruit trees are highly ornamental. They bloom in shades of pink, white and red in spring, come in sizes to fit any landscape situation and look handsome all summer. The British, who live in a climate too cold to get fruit, nevertheless plant peach trees for their ornamental qualities.

## Planting

While early spring is the traditional time to plant fruit trees, there is no reason they can't be planted through the summer as long as they are carefully watered. Garden centers and nurseries have a pretty good selection of fruit trees throughout spring, and they are often marked down later in the season. If you want to wait until fall to avoid the hassles of watering during summer dry spells, remember that you can safely plant apples and pears then, but cherries, plums and peaches should wait until next spring.

Another thing to remember is that you will need two different varieties of apples or plums to get a good crop; peaches, pears and cherries will usually produce well with just one of a kind in your yard.

You also have a choice of standard, semi-dwarf and dwarf fruit trees. Standard and semi-dwarf trees are the choices of professional fruit growers because they are most productive and somewhat less expensive than dwarf trees. But for backyard gardeners, dwarf trees make the most sense. Dwarf trees grow only slightly taller than a person, so the fruit is easy to reach and, more important, sprays can be applied effectively with equipment the homeowner can easily operate. You don't want to have to buy tractors and full-size sprayers to keep the insects off your three fruit trees.

You will go a long way to ensuring success with fruit trees if you plant them in the right place. Avoid planting fruit trees at the bottom of slopes where spring frosts will do the most damage. Plant in full sun and on sloping ground if possible. If you are planting several trees, be sure to know their ultimate height and width, and give them enough room so they are not growing into each other.

Dig a large hole for a small tree. I like to dig the hole out with a shovel, then use a stout digging fork to push holes farther into the soil to a depth of 3 feet or more. That allows for good drainage. If the tree is grafted—and most fruit trees are—set the tree in the hole no deeper than the bottom of the graft, which is a swollen place somewhat resembling a fist a few inches above the roots. (As the tree grows, cut away any suckers that come out from the base of the tree below the graft.) Some add fertilizer to the soil at planting; I wait until the next season to fertilize. Special tree fertilizers are available at garden centers, or you can add about half a pound of 5-10-10 the first couple of years and a pound or so after that. Apply fertilizer in a circle along the dripline of the tree.

## Pruning and care

When people say you can't grow fruit in Kentucky, often what they really mean is that you can't grow fruit that is picture-perfect, like what you see in the grocery store. There is a little more truth to that, because our climate here is conducive to insect populations and disease that attack fruit and cause cosmetic damage. There is a reason that much of the fruit you see in stores is grown in the upper Midwest or Northwest—insect and disease pressure there is minimal.

But you can keep insect and disease damage to a minimum.

The first line of defense is good site selection and fertility, mentioned above. The second is proper pruning. The mistake most people make is they don't prune trees severely enough, perhaps because they fear damaging the plant. Pruning is best done in late winter; February or early March, armed with a good book on pruning and a set of sharp shears, cut away. Apples and pears should be pruned along a central leader. Cut away all branches that grow toward the inside of the tree. Peaches, plums and cherries are pruned to a vase shape, with large branches coming out wagon-wheel style around the main trunk. But as with apples, keep the tree open by cutting out branches growing toward the interior. What you want to achieve (pardon me,

## RECOMMENDED VARIETIES

Your county Extension Service office will have a list of recommended fruit varieties for Kentucky. These are the varieties, starting in order of best, I have had the most luck with:

**Apples:** 'Enterprise,' 'Liberty,' 'Prima,' 'Gold Rush'

**Cherries:** 'Montmorency' or 'North Star' sour cherries

**Plums:** 'Green Gage,' 'Shiro,' 'Ozark'

**Peaches:** 'Red Haven,' 'Georgia Belle,' 'Elberta'

**Pears:** 'Kieffer,' 'Seckel,' 'Moonglow'

I have never grown nectarines, so I can't make a recommendation, but from all reports they are much harder to grow than peaches.

---

Humane Society) is a tree open enough that you can throw a cat through it. That allows for good air circulation, which goes a long way in controlling diseases.

If you do not spray your fruit trees at all, you will likely get some fruit if you follow good cultural practices. Apples and pears are most easily grown without spraying, especially if the right cultivars are selected. Peaches, cherries and plums are tougher to grow without spraying. But for better production of all fruits, a few well-timed sprays are important.

Fortunately, a number of organic insect and disease controls are available for backyard fruit production that do a good job of allowing you to grow quality fruit. Organic controls are somewhat more expensive than the chemical, one-poison-fits-all approach, but on the homestead level, that doesn't matter very much. Always read the labels carefully because even organic pest controls can harm the trees and the environment, especially bees, if applied at the wrong time or in the wrong quantity.

And learn that most insect and disease damage is cosmetic and certainly doesn't harm the quality of the fruit. In fact, when you grow your own fresh fruit, you can pick one off the tree and savor the juicy goodness, knowing that no one, no matter how much money he is willing to spend, can match the flavor and vitamin-filled quality you have right there in your hands.

# A flair for ferns

It is easy to get hooked on plants.

Aunt Sally gives you an orchid for Christmas and two years later you find yourself building a 900-square-foot greenhouse to support your addiction.

Or you buy a daylily at a discount store and a few years later you have a collection of 1,500 varieties and spend all summer hybridizing in hopes of producing the first-ever blue daylily.

Ralph Archer is hooked on ferns.

Archer, of Shelbyville, got hooked years ago when he bought a house in Middletown that included a very shady back yard. Ferns, natives of the woodlands, love shade, so Archer developed a love for ferns. Like all plant addicts, Archer is not satisfied with a few species. A retired quality control consultant, he kept adding to his collection until, at one point, he had as many as 75-80 species.

He joined fern societies and the Master Gardener program in Jefferson County. He propagated ferns. Encouraged by his wife, Jean, Archer went into the plant sale business, selling ferns at farmers markets. As his knowledge increased, he started writing articles about ferns. Archer's expertise on the subject is acknowledged by fern expert Sue Olsen in the introduction to her work, *Encyclopedia of Garden Ferns*.

When Archer and his wife decided to downsize, they moved into a patio home in Shelbyville. The home, in full sun near a golf course, was no place for ferns. So Archer donated the bulk of his collection to Whitehall, a historic home in Louisville built in 1855 on 11 acres of land on Lexington Road. (Whitehall is not to be confused with White Hall near Lexington, the home of Cassius Marcellus Clay.) Archer got involved with Whitehall when he was volunteering for his Master Gardener certification in Jefferson County. Now he volunteers there about three days a week. The shade garden there is named for him.

The Whitehall gardens, Archer said, have the largest fern collection between Birmingham, Ala., and Chicago, about 150 species. The ferns grow in a shady area of about three-fourths of an acre on the west side of the Greek Revival-style mansion. Archer and helpers cleared the

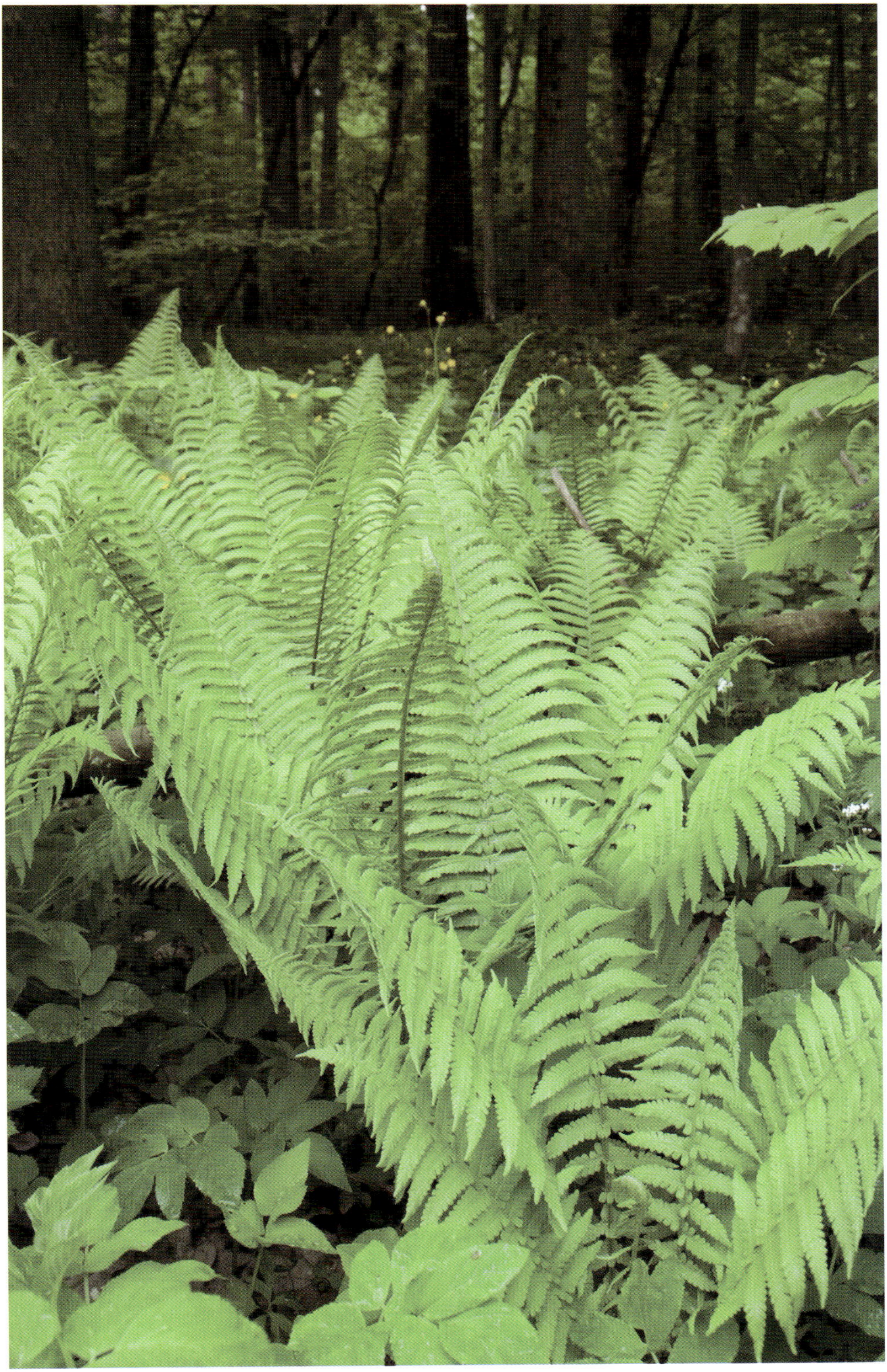

area, once choked with honeysuckle and ivy, by putting down flattened cardboard boxes to kill the weeds beneath. Then a tree care company put down about a foot of shredded trees and limbs. That set for a year, and then the ferns were planted directly into the mulch. Archer said the method is inexpensive, and the ferns seem to like it.

In the shade garden, Archer also built a stumpery, which he describes as "a poor man's rock garden." The stumpery mimics the appearance of a forest floor with ferns growing on, in and around stumps.

"We had lots of wood lying around, so we made use of it," Archer said.

Among the ferns, Archer and his crew also planted other shade-loving perennials, including Canadian ginger, astilbe, trillium and hostas.

## CARE AND FEEDING OF FERNS

Many fern species are native to the eastern woodlands of the U.S. Ferns everywhere are woodland plants that thrive in shade, though dense shade, such as that beneath a maple, should be avoided, Archer said.

Ferns are also adapted to woodsy soils, which are light and well-drained. Ferns do well in damp conditions as long as the soil is well-drained. A few species of bog ferns, however, can tolerate standing water.

To give the ferns the light, loose soil they desire, Archer mixes wood chips with soil and plants the ferns in the mix. In the stumpery, some ferns are planted directly into hardwood stumps that have a hole or hollow spot.

For the first year, ferns should be kept watered, Archer said. After that, they can withstand all but the longest dry spells. Fertilizer is usually not necessary, and most ferns have few, if any, insect or disease problems. Rabbits may nibble but rarely eat an entire fern, Archer said, and deer leave them alone.

"In the northeast, they have problems with snails, but around here, not much bothers them," Archer said.

## USE IN THE GARDEN

Ferns are planted primarily for their wide range of foliage shapes and forms, though the Japanese fern, with its silver and red leaves, is showy. Ferns come in every size from barely visible on the forest floor to huge specimens, taller than a man.

Ferns make a great backdrop plant for flowers that bloom in the shade, including goatsbeard, monkshood and astilbe. Other companion plants for ferns are hostas, toad lilies and hydrangea.

A stumpery is a great way to showcase ferns, Archer said, but he warns that it can be high maintenance. The stumps often rot and have to be replaced frequently.

## A GREAT MARRIAGE

The otherworldly fiddleheads of cinnamon fern open to set off the swanky beauty of astilbe. Plant them together in your shade garden.

# Create your own garden 'island'

I make no claim to being a creative garden designer. All of the good ideas I have used I've stolen from someone else.

I've visited a couple of city gardens that make creative use of what I call islands—groupings of plants surrounded by grass. Some of the gardeners used groupings of annuals, perennials and small trees in curvy beds that made their small yards seem larger than they were. Some used islands of tall shrubs and short perennials to create a quiet, shady nook for a bench. One gardener used an island to help hide an above-ground swimming pool. Another used a combination of small evergreens and perennials toward the front of the yard to give visitors a visual surprise—a whole other garden and stretch of lawn revealed when they rounded the back of the island. All of the gardens, with their islands of plants, had the effect of creating a quiet, green retreat in the center of the city.

Islands can be as simple as a couple of blooming shrubs combined for visual impact in the middle of the yard, a grouping of annuals—impatiens perhaps—surrounding the base of a tree, or a collection of hybrid tea roses set apart in their own bed. One of the most common islands shown in garden catalogs is a grouping of hostas surrounding a tree. The combination is effective because the hostas like the shade cast by the tree and the tree appreciates the large-leaved hostas cooling its roots and keeping string trimmers away from its base.

## Design

Islands are different than the more common annual and perennial borders in a couple of ways. Unlike borders, islands do not have a front, back and middle. The advice in the garden catalogs to "put this plant in the middle of the border" doesn't work as well for an island because the middle of an island is the "back," so to speak. And the back could be the front and the front the back, depending upon your perspective.

Another difference between the island and the border is that because the island is set "in the middle of things" as it were, scale, shape and visual perspective become more critical.

As far as scale, think about how the island will look in the scheme of your entire yard. A 4-foot by 4-foot island of low-growing annuals or perennials, for example, in a 3,000-square-foot backyard is going to get lost if not look downright silly. On the other hand, allowing the island to take up most of the yard is the other extreme. However, if yard space permits, you might consider planting an island of shrubs, trees and lower-growing perennials large enough to contain a path that leads to another part of the yard. In effect, you are creating two islands in your yard or even four if paths can crisscross.

Think about shape, too. Adjust the shape of the island to offset the shape of your yard. A square island in a square yard looks like you're playing with Lego blocks. In a square or rectangular yard, try a circular island. Better yet, create an island with an undulating pattern that lets the eye gently follow the curves and lines of the space. And don't feel the need to put the island in the middle of the yard. Offset it a little bit or put it toward the back or front, perhaps in a place where it might hide something, like that swimming pool or those trash bins.

Finally, think about placing the island where you might best enjoy looking at the plants growing there from the deck of your house or a kitchen or bedroom window. Some plants, magnolias or kousa dogwoods, for example, are best viewed from above.

## Plant selection

Before you start to choose plants that create a visual impact in your island, start with the basics—sun exposure, soil type, drainage, etc. If the island is in the middle of the yard, choose sun lovers; if sheltered by a line of trees, choose shade lovers. Remember that if you plant trees in the island, they eventually may provide shade of their own, so choose plants accordingly. By the way, the best trees to plant in an island are probably small ones. I visited one yard that had several islands of different plants but all were tied together visually by plantings of either Jane magnolias or dwarf river birch. Crabapples, Japanese maples or serviceberries are also good choices for island trees.

Remember, too, that large shrubs can give your island the height you want, and in less time than trees will.

## Building and Maintenance

Before you start hauling dirt or putting in plants, use a can of spray paint or even an old garden hose to outline your island. Kill the grass within the boundaries of the island—September is a good time to do that—by spraying with a total kill herbicide. At this point, some gardeners put down a weed-barrier fabric to block weeds, but I personally don't like the stuff; it invariably works itself through the mulch and after a couple of years weed seeds get a foothold anyway. And it makes later digging more difficult. Haul in enough topsoil to raise the island at least four inches above the surrounding lawn. Six inches would be even better because the soil will settle. But do not mound up soil near the base of the trees; that encourages rot.

When putting in plants, it's not a bad idea to crowd annuals and perennials together because as they fill in, they will reduce the need to weed; you can always thin them out later. Do not, however, overplant shrubs and trees or you will find yourself creating a tangled mess that some future homeowner will have to take out. If you don't like weeding, consider using a pre-emergent herbicide. Once the plants are full grown, they will crowd out most weeds. Occasionally, you might have to add some soil to the island as, over time, it will tend to sink. And there's nothing more disconcerting than a yard full of sunken islands.

## WASCALLY WABBIT

City gardeners have more problems with rabbits dining on their plants than we country gardeners do. If Brer Rabbit is pestering you, there are some perennial plants you can set out that will send him hopping off to the neighbor's. Some rabbit-proof perennials include: achillea, aconitum, digitalis, astilbe, bergenia, colchicum, geranium, gentiana, stacys, yucca, salvia, popaver, hosta, iris and hemerocallis.

# Summer

# Pass the veggies, hold the poisons

If you are going to grow food for your family's table—and I recommend you do—why not try to eliminate, or at least minimize, the use of chemical insecticides and fungicides?

In other words, go organic, or in today's parlance, "go green."

There is still considerable debate about whether vegetables and fruits grown organically are better for you nutritionally than those grown with chemical sprays. But there is little debate that organic controls are better for the environment, at least better for the environment in your backyard. Spraying kill-all chemicals harms not only the bug that's bugging your beans or tomatoes, it kills bugs that might be there to help you. And it may kill the birds that eat those bugs, the butterflies you want to attract, or the bees you want to pollinate.

Even though organic produce has gone mainstream and can be found in grocery stores from Whole Foods to Wal-Mart, too many gardeners still think that you can't produce fruits and vegetables here in Kentucky without chemical sprays. Nothing could be further from the truth.

The fact is, most vegetables and fruits can be grown most of the time without any chemical sprays whatsoever. Examples are beans, lettuce, peas, tomatoes, squash, okra, sweet potatoes, turnips, asparagus, spinach, beets, peppers, carrots and radishes.

A few may need spraying occasionally. Examples are corn, broccoli, cauliflower, cabbage, Brussels sprouts, Irish potatoes, blueberries, blackberries, kale and watermelons.

And there are some fruits and vegetables that are darn difficult to grow without using insecticides or fungicides.

Examples are cantaloupes, eggplant, and tree fruits. Fortunately, there are organic sprays that can be used on these (see box).

The key to gardening without chemical sprays is knowledge. Here is what you should know:

### Know your soil

"Feed the soil and the soil feeds you" is the motto of the organic gardening movement. Just as malnourished humans are more prone to disease than well-fed people, plants grown in malnourished soil will fall prey to insects and disease far faster than those grown in rich, organic soils. Have your soil tested by your county Extension Service at least every couple of years and add soil amendments to beef it up when necessary. Compost is an excellent organic fertilizer; you can buy it if you can't make it. Animal manures, cover crops, spoiled hay, even leaves, added to the soil improve both fertility and structure.

### Know your varieties

Plant breeders are constantly creating vegetables and fruits that resist insect invasion and disease. If a variety of tomato, for example, does not do well for you, try another one. Look for "disease resistant" or "insect resistant" on the package before you buy. But while trying something new, don't neglect the old varieties. Many of the heirloom vegetables and fruits grown before the advent of pesticides have more built-in resistance than modern varieties.

### Know your friends

Few insect species flying and floating around your garden actually do any harm. And some are out there helping you by taking out your enemies. Help those beneficial insects by planting flowers that attract them. Examples: Queen Anne's lace, yarrow, alyssum, tansy and clover. You also help those beneficial insects when you avoid spraying total-kill insecticides.

## SPRAY ORGANIC

Organic sprays and controls are being created every day to support the growing demand for organic produce. Here are a few you might use in the home garden:

**Neem** - an insecticide and fungicide in one package, Neem is derived from the leaves of a tropical Asian tree. Effective against soft- and hard-bodied insects.

**Bt** - short for bacillus thurigiensis, Bt gives caterpillars a stomachache; they stop feeding on your corn, cabbage, cauliflower and broccoli and die within a few days. Avoid Bt if you want to attract butterflies.

**Pyrethrins** - A quick knockdown insecticide derived from the flowers of a tropical daisy. Effective against chewing insects and beetles.

**Rotenone** - Derived from the root of a tropical plant, rotenone is deadly to a wide range of insects but less toxic to warm-blooded animals. Use as a last resort against beetles and other pests. Be careful not to inhale rotenone.

## Know your enemies

Knowing what's doing the damage to your plants is half the battle to controlling it. If it is the ravages of disease, it is likely too late to spray. Fungicides must be on the plant before diseases get a foothold. So hold the fungicide and vow to try another variety that may resist the disease.

If an insect is munching on your tomatoes or worming its way into your corn, first identify it, then learn the least harmful way of controlling it. Two good books that help with both insect identification and control are *The Organic Gardener's Handbook of Natural Insect and Disease Control* put out by Rodale Garden Books and *Grow Organic* by Doug Oster and Jessica Wallisher.

What both these books will tell you is that sometimes spraying is necessary and sometimes letting nature take its course is better than intervening. For example, tomatoes are often afflicted by a giant green caterpillar called a hornworm. You can kill them with a spray, but if you find a hornworm with little white specks on its back, know that a tiny wasp has laid its eggs on the worm; the eggs will hatch and the larvae will eat the caterpillar. Gross, but effective control. And the larvae will produce more wasps that keep the caterpillars under control.

Finally, if you are going to grow organic, you will have to accept that some of your vegetables and fruits will not be blemish-free, but they are edible nonetheless. That picture-perfect fruit in the grocery store comes at large cost to our environment and perhaps to our health, a price we don't need to pay if we grow our own without poisons.

## HERBS FOR THE GOOD GUYS

Planting herbs among the vegetables not only allows you to collect your veggies and their accompanying spices at the same time, it attracts beneficial insects that prey on the bad insects. Dill is one of the best at attracting beneficials. Yarrow, feverfew, sage and thyme are other good hosts for beneficial insects.

# Let the sunflowers shine

The all-American sunflower is now on its third career.

For centuries, the sunflower was a source of oil—for Native Americans and later the colonists. In fact, evidence suggests Native Americans cultivated sunflowers before they grew the "three sisters"—corn, beans and squash. Sunflower oil is still commonly used for cooking and as an ingredient in biodiesel. Most oil-seed sunflower is grown in the upper Midwest.

In the last century, people started feeding birds to attract them to their urban and suburban yards, and they discovered that nearly every seed-eating species relishes sunflowers, especially the black-seeded varieties. And people learned that, roasted and salted, sunflowers make a pretty good snack. Though high in fat, sunflower seeds are rich in minerals, including calcium.

Within the last decade or so, plant breeders have turned the sunflower into an ornamental with a wide range of possibilities in the garden. The big-headed varieties are still around, but sunflowers have been bred to be no taller than 18-20 inches, fitting perfectly in pots or blend into the front or middle of a sunny border. In 2004, the Fleuroselect organization gave a quality award to the "Pacino Cola" variety, which produces large flowers on dwarf plants, perfect for containers. Pair container varieties with a sun-loving, cascading plant, such as 'Purple Wave' petunia and you have a winning combination.

Sunflowers have also been bred in double form. The 'Teddy Bear' variety is a popular double that looks cute enough to cuddle up with. It, too, works well in containers and pots as well as the front or middle of the border.

Also, sunflower lovers have found heirloom and old-fashioned varieties that come in dozens of colors—brilliant orange, red and maroon, creamy white and multicolored. Some of the oldies are branching plants that have dozens of smaller seed heads rather than the one large "sun." Tall plants with scraggly looking bottoms, these multicolored sunflowers look best in the back of the border, shining behind summer annuals or perennials such as phlox, rudbekia, roses or summer lilies. The tallest types would go well in back of a butterfly bush or hardy hibiscus.

## CULTURE

Sunflowers (Helianthus annuus) are native to much of the U.S. Before people came along to cultivate them, they grew in openings in the forest and in the prairies.

Sunflowers like sun, and they like it hot. Though they need some moisture to get established, sunflowers can take relatively dry conditions. So do not water your sunflowers unless the summer turns unusually dry.

Plant sunflowers in late May or early June for blooms in August through early October. Plants bought at garden centers may already be in bloom, though for best quality, choose those that have not bloomed yet.

Sunflowers for ornamentals can be planted about a foot apart. Those planted for seeds, the large-flowered varieties, should be planted at least 2 feet apart. Large-flowered types may also be staked once they start to bloom if the garden is subject to wind damage. Once the plants are up, mulch them to keep down weeds. Sunflowers are bothered by few insects and diseases if they are grown in full sun. As long as sunflowers are planted in reasonably fertile soil, they do not need fertilizer. If you do fertilize, go with a product low in nitrogen, such as 5-10-10, and apply no more than half a cup per plant about the time of bud formation.

The sunflower's Latin name

## A SELECTION OF SUNFLOWERS

**'Autumn Beauty'** - 7-foot plants with red, gold, yellow and rust colors.

**'Lemon Queen'** - Lemon yellow flowers and a deep brown head

**'Italian White'** - 4-inch creamy flowers with a chocolate brown center

**'Teddy Bear'** - 3-6-inch double flowers on dwarf plants

**'Velvet Queen'** - 5-inch maroon blooms

**'Evening Sun'** - 7-foot plants with bicolored flowers

**'Black Russian'** - The large-seeded variety for birds

**'Mammoth Russian'** - A large-seeded, confectionary type

**'Torch'** - Fiery orange flowers on 4-to-6-foot branching plants

**'Valentine'** - Yellow tips on petals get darker as they move to the center. One of the best for cut flowers

"helianthus" means sun follower, and that is exactly what sunflowers do when they are in the bud stage. The buds will face east in the morning and gradually turn to the west as the day progresses. Many flowers exhibit this trait, called heliotropism. Once the sunflower starts to bloom, the sun chasing stops because the stem grows stiff. Flowers at that point will usually face east.

Some sunflowers make great cut flowers that last for days in a vase. The seeded varieties are magnets for the brilliant yellow and black goldfinches that will perch upside down for a snack.

If you want some seeds for yourself, harvest the plant once the seeds have fully matured (the outside petals will have wilted away) on a warm, dry day. Cut at least a foot of stem along with the head and use that to hang it in a warm, dry place. Put a paper bag under the head to catch those seeds that will fall as the plant dries.

## SUNFLOWER POLES

Sunflowers are attractive in the vegetable or flower garden. Make the stalks of giant varieties do double duty. Plant pole or runner beans at the base of the stalks and let them climb to the top. Because sunflowers release a chemical that inhibits germination, you might want to try pre-sprouting the bean seeds first.

# Gardening inside and out

With baby boomers aging and gas prices jumping, more and more Americans are leaving that chunk of land in the suburbs and getting smaller digs. Some are choosing apartments, some condos or townhouses, others small city lots.

That does not mean they have to give up gardening. They may just be measuring in square inches rather than square feet.

Bringing the garden inside, or at least closer to the back porch, presents challenges but opportunities as well. Gardening on a small scale—in containers on a windowsill, in pots in a Florida room, or on a tiny strip of land between sidewalks—requires careful selection of plants, close attention to cultural requirements and timely care. Plants in containers are like animals in a cage; they can't go foraging for their needs in the surrounding soil—you have to provide them. On the other hand, if their cultural requirements are met, nearly any plant, from a fruit tree to a tomato to the makings of a salad to ornamentals can be grown in a small space inside the house or just outside the door.

## THE PLANTS

Plant breeders have produced dozens of trees, shrubs, vegetables and ornamentals that fit well into small spaces and containers. If you have room for a 5-gallon container on your deck or patio, you may want to try growing one of the apple trees bred to grow straight up rather than out. When in fruit, the trees look like a stick covered with apples. Tiny peach, apricot and nectarine trees have been bred to grow in containers. Lemon and lime trees have been grown inside Florida rooms, attached greenhouses, sun rooms, even living rooms, for years, though they are best taken outside during the summer.

Dozens of dwarf shrubs are good choices for small spaces on the patio or deck. Look for dwarf itea, clethra 'Hummingbird,' fothergilla gardenii, dwarf conifers and boxwood.

Most people, when they think of indoor plants, think of ornamentals only. But in fact, vegetables and herbs can be grown inside. Cherry tomatoes will grow in a small space. Regular-size tomatoes can be grown in a container if you have a way of staking or supporting them. Some inside gardeners surround the containers with wire to hold the plants upright. Bush varieties of cucumbers need very little room. Add a container full of lettuce and you have your salad fixin's. When choosing vegetable varieties for the kitchen garden, if that garden may be literally in the kitchen, look for names such as "container" or "patio."

Be warned that if you grow any fruit or vegetable inside, you will have to perform the pollination for the fruit that the bees and flying insects would do naturally outside. It is merely a matter of using a small, soft-bristled brush and touching from one bloom to another. But if that is too much plant sex for your taste, you might want to stick to vegetables where the leaf is the edible part, such as greens and herbs.

And don't forget to pile on the ornamentals, whether the garden is in the kitchen or on the deck. In fact, mixing ornamentals with vegetables produces some interesting and colorful combinations. Try growing yellow marigolds with purple basil and chives. Combine tomatoes with red-plumed celosia or grow broccoli in a container with cascading petunias as a colorful understory.

## Cultural requirements

When growing plants in containers or in small spaces, focus on four requirements: light, moisture, drainage and food.

Plants grown indoors or just outside will have the same need for light that plants grown in the yard have. You just may have to be more watchful of where you put what. Vegetables, herbs and fruits need full sun, which means they will grow inside in a sun room or Florida room but are going to underperform in a dimly lit room on the north or east side of the house. If you have no "full sun" rooms, you can use grow lights and stands to give them the growing lumens they require. Even in a well-lit room, plants will tend to grow toward the most light, so you may have to turn them around occasionally to keep them from growing lopsided.

Plants in containers, sitting on decks or hanging from hooks, will need more constant monitoring of their moisture needs. Plants in containers, especially large vegetable plants and small trees and shrubs, dry out quickly and some may need watering as often as twice a day. Having said that, more plants are killed by overwatering than underwatering. Do not water every time the surface is dry. Scratch the surface and see if the soil is moist about a half-inch deep. If damp, hold the water. If dry, water until it runs out of the drainage hole.

Which brings up the next need: drainage. A plant's roots are both its stomach and its lungs. It can't breathe if saturated with water. So a container of any size must be freely draining. No matter how large the container, if it does not have drainage holes, the plant will eventually suffer root rot from sitting in standing water.

Finally, feeding; Plants grown in containers can be much trickier to fertilize than those outside where the plant's roots can compensate somewhat for your lack of knowledge by moving into more hospitable turf. Most experts recommend some form of slow-release fertilizer for container-grown plants to avoid the feast-or-famine syndrome. Slow-release fertilizer is incorporated into some potting soils or can be bought as spikes or tablets or added to water in small amounts.

Start small, add a new plant a year and turn your house into a greenhouse.

## KINKY CONTAINERS

Terra-cotta pots and their plastic counterparts are fine containers for growing indoors or just outside. But when it comes to containers for your plants, don't limit yourself to the obvious. The only rule is that the container must hold soil and have holes for drainage.

Get a kid's red or yellow rubber boot, fill with soil and plant with purple and white petunias.

Plant herbs such as sage, basil and thyme in an old wheelbarrow sitting by the doorstep. Grow chives in a gallon coffee can.

Or make a plant bed—literally. Use an old metal bed frame, sit it next to the door and plant cherry tomatoes inside. Grow a cucumber up the backboard.

## CHEERS!

Plant-sucking insects such as scale do considerable harm to inside plants. And spraying insecticides inside the house isn't a good idea. Try dipping a cloth in rubbing alcohol and rubbing the infected leaves. The alcohol kills the scale and seems to keep them away for a long time. Apparently they're teetotalers.

# Don't slime okra

Okra is the Rodney Dangerfield of vegetables. It doesn't get any respect.

Perhaps it is because the vegetable was associated with the slave trade. Okra originated in Africa, and the vegetable came to the U.S. via the West Indies with slave vessels in the 1600s.

Then there are the prickles. The fruit in old varieties of okra is protected by prickly barbs that can poke and irritate the hands. More modern varieties are less prickly but still irritate sensitive skin. Picking with gloves is never a bad idea.

And then there is the slime. Cooked certain ways, okra releases a clear slime that can be off-putting. But if you grew up Southern and know how to cook, you know how to fix okra so that it is not slimy (see recipe) or you know how to use the sliminess to advantage—to thicken soups, stews and gumbos. In fact, okra is the traditional main ingredient of Louisiana gumbo. The word gumbo comes from the West African word for okra.

Even if you don't cook it, okra is worth growing as an ornamental. Okra is a tropical plant, related to cotton and hibiscus. Blooms are similar in size and shape to the blooms of rose of Sharon but are usually yellow. Red okras have a bloom that is rose-colored. Okra blooms happily in the heat of the summer when other plants are in decline and will bloom until frost if you keep the pods picked.

It grows tall, with large, fan-shaped leaves offering an exotic look to the garden. Because the plant has few or no bottom leaves, okra is a good candidate for planting among lower-growing annuals and perennials. Okra would look good behind peonies, for example, or among daylilies, marigolds and zinnias. I once saw okra growing among summer phlox, an interesting combination of colors and textures.

The pods of okra are also interesting and very durable. Decorators use the spear-shaped pods, especially the red ones, in dried arrangements.

## Growing okra

For years, most fans of okra could find only one variety to choose from—'Clemson spineless.' 'Clemson spineless' was a breakthrough in okra breeding back in the 1930s because it was, as the name suggests, less painful to pick than the older varieties. It also stayed more tender on the plant longer than older varieties. But "spineless" is something of a misnomer. Picking 'Clemson spineless' can still be somewhat irritating.

In recent years, with the rise in popularity of Cajun cooking, plant breeders have turned their attention to creating more varieties of okra. Dwarf varieties—plants growing no more than 3-4 feet tall—are now common. Try 'Lee,' 'Annie Oakley' or 'Cajun.' 'Burgundy' is a modern, dwarf red variety; many gardeners like it as much for the red color of the pod and flower as for the fruit.

Some of the older varieties of okra are also interesting. 'Cow Horn' is an heirloom from the 1800s that has pods 6-8 inches long. 'Star of David' has chunky, star-shaped fruit. 'White Velvet,' 'Red Velvet' and 'Green Velvet' are old varieties that look just like their names suggest.

Whatever variety you choose, okra likes it hot. If you planted okra in early May, you were likely disappointed. Okra will not germinate well until ground temperatures reach nearly 75 degrees. Late May through mid-June is a good time to plant.

Plant okra in moderately rich soil that has been worked to a fine consistency. Space seeds about 6 inches apart. Okra can be slow to germinate; be patient. Eventually, the seedlings will appear, looking like little green umbrellas.

## FREEZING OKRA

Okra is an excellent vegetable to store frozen. I cut off the stems and tips, slice into approximately half-inch pieces, rinse in a colander, drain and store in pint-size freezer bags. Add okra to winter soups and gumbos to add thickness and texture.

## FRIED OKRA

20-25 fresh okra pods, 3-4 inches long

2 tablespoons flour

2 tablespoons corn meal

1 cup milk

1 egg

1 teaspoon salt

1 teaspoon black pepper

Cut both ends from okra and slice into pieces about a quarter-inch long. Rinse in a colander and drain. Beat egg into milk and pour over okra. Mix and pour off excess liquid. Mix flour, cornmeal, salt and pepper and pour over okra. Coat okra well. Heat about a half-inch of oil in an iron skillet to hot but not smoking. Pour in okra and spread out. Fry for about five minutes and flip pieces. Drain on a paper towel.

---

Keep weeds down until the plants get up to about 6 inches; then okra will outgrow anything underneath it. I like to hill up the plants with a few inches of soil around the base to keep the taller varieties from blowing over in a windstorm.

Except for an occasional attack by Japanese beetles, okra is seldom bothered by insects or disease. Once you get okra out of the ground, you've got it made.

Okra will bloom when the plants are as small as 6-8 inches. Pods will start forming shortly thereafter. Now the work starts. Okra is extremely productive, so I hope you have planted no more than 10 or 20 feet of row. Okra should be picked every two to three days when the pods are no more than 3 inches long. Wait too long and the pods get tough and woody. Be warned that you need to go through the okra patch carefully; the pods can be hard to see. I have picked through a row only to go back and pick almost as much as I did the first time.

If you have an excess of okra—and you will—it does freeze well for later adding to soups and stews. No special processing is needed.

Pretty versatile for a vegetable that gets no respect.

# Try squash without the guilt

Summer squash—zucchini in particular—probably has inspired more garden guilt than any other single plant.

The guilt stems from the ability of zucchini to produce bushel after bushel of fruits even after the freezer is full of shredded zucchini, zucchini muffins and zucchini bread. It keeps producing loads of little green cylinders even after the neighbors start shutting their doors and pulling down the shades when they see you coming with another "gift from your garden." And we gardeners just can't let anything go to waste.

Nor can you ignore zucchini. Stop picking the fruits and zucchini just will keep right on growing to the point your garden looks like a group of Neanderthal men left their clubs lying about.

Many gardeners stop growing summer squash because of the zucchini flood and subsequent guilt, and that's too bad. All squash—summer, as well as winter squash—are full of vitamins and deserve a place, even if it's a small space, in the vegetable garden. Native Americans, who knew a thing or two about gardening, grew squash as one of their "three sisters" of food staples. Beans and corn were the other two "sisters," and they often were grown together, with the beans growing up the corn and the squash covering the ground to conserve moisture and discourage marauding raccoons.

You just need to learn to manage the flood, that's all. Instead of planting the entire seed packet at one time, plant one or two plants every couple of weeks throughout the summer, ending by the middle of July. I like to plant one green zucchini and one yellow summer squash at a time. That gives me a fresh and steady—and manageable—supply of summer squash until frost.

## SUMMER AND WINTER

Basically squash come in two kinds—summer and winter. The summer squash are the zucchini, usually green; the yellow squash; and the patty pans (scallop). Scallops are a more densely textured summer squash, but yellow squash and zucchini are similar in taste and texture, although gourmets claim to be able to detect subtle differences in flavor between the yellow squash and zucchini. Both are used raw for dips and in salads or, in the case of zucchini, as the key ingredient in breads and muffins. The greens and yellows look great together on the plate and in stir-frys, which is why I like to plant one of each.

Summer squash can be large plants, but if space in the garden is at a premium, try some of the varieties that are labeled "space savers" or "compact bush." They will grow full-size fruits on smaller plants. Some summer squash varieties even will do well in containers on the deck or patio. Just be sure to give them full sun.

Winter squash come in a dizzying array of varieties: acorn, butternut, banana, hubbard, delicious, pumpkins and spaghetti. Winter squash are planted less frequently than summer squash, probably because they are more inclined to grow on rampant vines and take longer to mature. Winter squash take at least 75 days to mature and many require as much as four full months of growing. If you want winter squash to mature before frost, you need to get them planted by the end of June at the latest.

Unlike summer squash, winter squash produce a more controlled harvest of fruits, but a few hills will still produce plenty for an average family. Most winter squash also store reasonably well; butternuts, for example, will keep up to five months in a cool, dry environment. Winter squash are chock full of vitamins, especially vitamin A.

## CULTURE

Because squash are very productive, they need a rich, well-drained soil. And because they like heat, always plant them in full sun.

I like to plant summer squash in hills about 3-feet-by-3-feet apart. A hill, in vegetable planting parlance, is not a mound but rather a grouping of seeds. Before I plant each hill, I add a shovelful of manure or compost and lightly cover with garden soil. Atop the garden soil I place four or five seeds and cover with about an inch of dirt. Once the plants pop through the ground, I thin to no more than three per hill. Crowded plants invite insects and disease.

Squash are hardy and will withstand some neglect, but if extremely dry weather hits, water the hills deeply at least once a week. If you grow squash in containers, you'll likely need to water them every day, especially when fruit production begins.

Squash can develop a number of diseases, including powdery mildew (looks like white powder on the plant), anthracnose, downy mildew and mosaic. Fortunately, diseases are rare because most modern varieties have been bred to resist them. If a plant does get diseased and dies back, immediately pull it out of the garden and dispose of it. Because many diseases are soil borne, make note of where the plant grew and do not plant any of the cucurbit family (including cantaloupes and cucumbers) in that spot for several years.

I have more trouble with insects than disease. Squash vine borers cause the plant to suddenly wilt and die. The borers burrow into the base of the vine and feed, cutting off the supply of water. Dusting the base of the plant with rotenone (if you use organic methods) or Sevin dust will kill off vine borers. Try to keep the dusts away from the blossoms, because you don't want to kill the bees that are visiting and pollinating your fruits.

Squash bugs, commonly called stinkbugs, and yellow striped cucumber beetles will also hit squash and suck enough juice from the vines to harm the plants. Cucumber beetles also spread a disease called cucumber wilt. The dust will keep those bugs at bay, but I find both species are rarely a problem early in the summer and more likely to hit from late July through August. A more benign approach to controlling those bugs is to turn the squash leaves over and find and eliminate the eggs before they hatch. The eggs of the squash bug look like small red dots; the eggs of the cucumber beetle are usually yellow.

Squash 'em!

## TRAP 'EM

Cucumber beetles are one of the worst enemies of squash because they transmit disease that can eventually kill the vines. Insecticides are difficult to use around squash because the bees are continually working the blooms and you don't want to harm bees. Garden centers now sell yellow traps that are supposed to catch cucumber beetles on a sticky pad. Haven't tried them and can't vouch for them, but at the very least the traps will let you know that the beetles are around and you can take retaliatory action.

# Q&A: What's bugging you this summer?

I'll take dealing with insects in the garden any day over fighting weeds, but a lot of gardeners throw in the "trowel" this time of year because they get tired of fighting bugs. They don't like little green worms bedded down in their broccoli, sweet corn chewed on by little white maggots, tomatoes gouged out by hornworms, or lawns yellowed by chinch bugs. Imagine that!

Fighting insects over your produce, lawns and flowers can be a challenge. I get lots of questions about insect control in summer. Here are some general rules to play by:

Remember that most insects are harmless and many are beneficial, so there is usually no reason to use a "general purpose" insecticide that kills every crawling, creeping and flying thing out there. You're actually likely to do more harm than good because the bad bugs eventually get immune to your poisons, and the good ones go away. Know your enemy before you spray. Your county Extension Service office is a good source of bug identification assistance.

Much of the damage insects do is cosmetic. That's a problem in the flower garden and perennial bed, but not so much in the vegetable garden. A gardener once told me that if he had to spray so much poison on something that an insect won't eat it, he didn't want to eat it, either.

Try organic insect controls first. They are readily available in garden centers and cost little, if any, more than chemical controls.

If you do use chemical controls, read the label carefully and match the chemical to the problem. Also, use the recommended dosage; the goal is killing, not nuclear annihilation.

**Q:** Some of the leaves of my perennials have yellow, speckled areas and drop off; the plants are often stunted. Sometimes I see small webs on the plants.

**A:** It's likely you are dealing with spider mites. They are especially troublesome in dry weather and seem to go away when rains come. You can buy oil sprays that smother the insects or you can use soap sprays that will kill the mites but do little harm to the environment.

**Q: Something is eating my tomato plants and even taking bites out of the green tomatoes.**

**A:** It's probably the tomato hornworm, a large green worm that looks like it was designed by Satan himself. If you see entire leaves eaten down to the stem, you can bet it's a hornworm. Look closely; you'll see him lying aside a stem. Pick him off. He won't bite; he just looks nasty. Alternately, you can spray with a product that contains Bt, a bioinsecticide that targets only caterpillars. But if you have just a few plants, I would control by hand.

**Q: About this time of year, my cucumbers and cantaloupes wilt and die suddenly just as the fruits start to mature. How can I avoid this?**

**A:** What you're seeing is the result of feeding by the striped or spotted cucumber beetle. They feed on cucurbits—squash, cucumbers, muskmelons—but also on roses and corn. Their eating does little harm, but they pass along a disease called bacterial wilt that plugs up the stems of cucurbits and causes them to suddenly wilt and die. The damage looks like someone took a blowtorch to the patch. This is one time you might consider a chemical control—malathion is a good choice because it kills the beetles but does little harm to bees needed for pollination. Still, spray according to directions and in the evening or early morning when bees are not out.

## WHAT A WAY TO GO

If you garden organically, you likely do a lot of hand picking of insects. You might consider keeping a Mason jar in the garden shed with a little rubbing alcohol in the bottom to serve as an execution method for bad bugs. Just drop them in the jar as you wander through the garden, and they'll be headed for nirvana in no time.

**Q: I'm hearing news about an insect called the emerald ash borer. Will it really kill my ash trees, and is there anything I can do to stop it?**

**A:** The emerald ash borer was first found in Kentucky in 2009 and will likely kill millions of the state's ash trees if the experience of our neighboring states that have already had the borer for years holds true. The borer can be stopped if you are willing to spend several bucks per tree fending it off. Insecticide drenches are available (brand name Bayer Advanced) that can be sprinkled around the tree and watered in. Start control in early spring and follow label directions.

**Q: My lawn gets yellow, dead patches every year and the grass gets thin. What can I do to prevent that?**

**A:** Could be several insects at work, including chinch bugs, grubs and sometimes sod webworms. Webworms and chinch bugs are best kept at bay by keeping the soil moist. Grubs can be controlled by using a product called milky spore, which gives them a disease. It does take a few years to work, however. The best defense against insects in the lawn is to use recommended turf varieties, fertilize in the fall and keep the grass mowed at a height of 3 to 4 inches. Many lawn problems are caused by scalping the grass when mowing, over-fertilizing or planting grasses that just do not hold up well in Kentucky. Again, your local Extension Office Service is an excellent source of help for lawn maintenance.

**Q: Is there anything I can do to keep Japanese beetles off my roses?**

**A:** Japanese beetles should disappear by late summer. They are here from approximately late June until mid-August, so you should start to see some relief. Their damage to roses is primarily cosmetic, but there's nothing so disconcerting as seeing a bud of your beautiful Peace rose covered in an orgy of metallic beetles that have no regard for what they do in public! Sevin dust certainly kills them, as do sprays containing imidacloprid or malathion; the latter is preferable because it does less harm to bees. You can pick them off in the morning when they're sluggish and drop them in a bucket of soapy water to drown. Try not to succumb to buying a Japanese beetle trap. Yes, you'll get beetles falling into the trap—and they'll be everywhere else in your yard as well.

# Rein in your garden bullies

For most of us, gardening is about nurturing. We fight the everlasting battle against weeds, insects, lousy soil, or bad weather that would keep our plants from growing. But sometimes we find ourselves with plants that, once established, not only thrive without our care, they actually take over the garden.

These invasive plants sprout where they don't belong, crowd out their neighbors, overwhelm a color scheme and sometimes tear up garden hardware.

## INVADERS

Everyone is fond of the herb mint, and many decide they will grow their own mint julep garnish in the garden. What they don't realize is that once you plant mint, you will always have mint. And not always where you want it. Mint spreads by underground runners. It will quickly colonize into nearby plants, pushing up into your daylilies, iris and peonies. It will smother lower-growing herbs. It will sprout in your yard (smells good when you cut the grass, though). Mint is best planted in a container, where its roots are confined. Just be sure to provide adequate moisture and good drainage.

Another notorious spreader is bamboo. Gardeners are warned to avoid planting bamboo unless they are willing to battle its spreading shoots for eternity. Bamboo has the reputation of being able to spread its roots beyond an 8-foot concrete barrier, push aside rocks and find its way through brick pavement. In warmer climates, bamboo has smothered entire lawns.

Like mint, bamboo is best planted only where it can be contained. And containing bamboo requires surrounding it by aluminum or some other metal barrier planted at least 18 inches deep. The other option is to plant only the clumping, as opposed to running, bamboo types. Those still spread, but not as vigorously.

While bamboo and mint invade from underground, many overly-vigorous garden plants spread from above, by scattering hundreds or thousands of seeds throughout the garden. The coneflower is a good example. It is a popular garden plant because it is a native of the prairie, it blooms most of the summer and it is little troubled by insects or disease. But coneflower is a prolific seeder. It will send up its babies, borne of seeds scattered by wind and birds, throughout the garden. You may have put your coneflower in the back of the border, but in a year or two it will march to the front, where it will flop over walks and overwhelm lower-growing specimens. If you want coneflowers but do not want to battle the seedlings, you may try some of the newer hybrids that do not reproduce themselves, or at least not as prolifically. Another trick is to plant coneflowers in a bed by themselves where their replenishing themselves by seed is welcome.

Garden plants with too stout a constitution sometimes tear up more than the garden.

Many gardeners have brought home little sprigs of wisteria because they admired its strings of purple flowers hanging from the vines in spring. They may have visited Biltmore in North Carolina and sat on the veranda beneath the shade of the wisteria winding its way above. But what they may have failed to notice is the stout woodwork the wisteria is clinging to. Plant wisteria next to your home or garage and within a few years it may crawl into the attic, pull down gutters, rip up the roof and pull windows aways from their casing. Wisteria is best planted on a sturdy arbor away from a dwelling.

## IN THE RIGHT PLACE

These extra-vigorous garden plants are not bad plants. Far from it. Like the playground bully, they just need a little careful watching and their energies channeled appropriately.

I once wrote a column praising the virtues of an old-fashioned, but now

---

## GARDEN PLANTS THAT CAN TAKE OVER:

**Perennials**—Mints, tansy, gooseneck loosestrife, obedient plant, creeping Jennie, Bishop's weed and sages.

**Grasses**—Horsetail, Japanese blood grass, dwarf ribbon grass, bamboo and prairie cord grass.

**Vines**—English ivy, wisteria (American and Chinese) and bittersweet.

**Shrubs**—Chokecherry, barberry, yews, glossy abelia, boxwood.

**Groundcovers**—Wintercreeper and akebia.

**Trees**—Mimosa, black locust and maples, especially red and silver.

seldom planted, relative of the hollyhock called malva or mallow. Malva is lower-growing than hollyhocks, has handsome striped flowers that are borne all summer on foliage that stays fresh and green, and avoids the hollyhocks' tendency to get ratty. After that column, a writer scolded me for praising a plant that multiplies like rabbits. It's true. Malva, like its cousins the hollyhocks, or the aforementioned coneflowers, scatters its seeds by the thousands. I have malva coming up yards from where it was planted. Some plants have even come up on the opposite side of the house. Malva babies sprout between the stones in the garden edging and manage to gain a foothold in a crack in the concrete sidewalk.

But sometimes the effect is marvelous, and the malva combines particularly well with a plant I had not thought of planting it near. Sometimes the malva brightens up a spot in the garden where the plants I had deliberately put in died.

So, as I told my critic, malva is a good plant, but just be careful where you put it, and then be willing to weed it out when it gets out of control. It's the same gardening advice you will hear over and over: Put the right plant in the right place.

## KNOW YOUR BAMBOO

Bamboo is notorious for being one of the most aggressive growers on the planet. Many a gardener has planted bamboo to provide an attractive screen or to grow where nothing else grows only to have to move out of house and home a few years later because the bamboo spreads uncontrollably. It is very difficult to contain, so if you must have bamboo, choose the clumping, rather than the running varieties.

# Sage gardeners plant salvia

It is not often that plants make the news, but the perennial herb salvia divinorum has gained attention, or notoriety, to salvia, a great family of garden plants.

You see, some adolescents and young adults have discovered that the herb, native to Mexico and usually called salvia on the streets, has psychoactive and hallucinogenic properties when smoked or chewed. Psychotropic effects include perceptions of bright light, hallucinations, a sense of loss of body and overlapping realities.

It is good that the salvia family is being discovered; it is a shame it is for the wrong reason. The salvia family, relatives of mint, contains dozens of species and hundreds of varieties of annuals, perennials and small shrubby plants that ought to be more familiar to gardeners. Though most are at home far south of here, salvia species include a range of plants we can use to great effect in Kentucky gardens.

Salvias are tough plants that thrive in hot, dry weather. They are seldom bothered by insects and disease. Butterflies and hummingbirds will come a-callin' if you plant salvia.

## The salvia family

Perhaps the most familiar salvia is the culinary herb we call sage. Sage is a prime ingredient in many sausage recipes, flavored dressings and herb vinegar. Common garden sage, Salvia officinalis, is a low-growing, silver-green herb with blue or lavender flower spikes and leaves with a pebbly effect. Some varieties have purple or golden variegated leaves.

While sage is most commonly used for cooking, it has also been considered a medicinal herb since ancient times. Some scientists today think the ancients were on to something; sage is being investigated for its natural estrogen and antioxidant properties. Sage tea can be made by steeping one teaspoon of dried leaves in a cup of hot water for 10 minutes.

Two salvia species, Salvia splendens and salvia farinacea, are tender perennials that we in Kentucky plant as annuals. Because they bloom all summer, salvias should be available in garden centers across the state into late summer. Often called scarlet sage, these salvias actually come in a range of colors from deep red through white, yellow, pink and purple.

My favorite salvias are the hardy perennials, Salvia superba and Salvia nemerosa. German nurserymen have been developing perennial salvias long before Americans and many of the best types have German names, including 'Mainacht' ('May Night'), and 'East Friesland.' 'May Night' was the 1997 Perennial Plant of the Year. Americans are starting to catch up, and dozens of varieties are being developed every year.

Perennial salvias usually have deep green, pebbly leaves and produce purple, pink, blue or white flower spikes on plants 18-36 inches tall. Blue and purple are the dominant colors.

Clary sage, Salvia sclarea, is another fine plant, though it may be more difficult to find than the perennial sages. The flowers are white with lilac markings. Clary sage has medicinal properties and has been used commercially as an additive to tobacco.

Garden writer Jeneen Wiche introduced me to another very hardy sage, blue meadow clary, Salvia pratensis. It produces dark green, pebbly leaves that stay close to the ground. It has long flower spikes with a shorter bloom period than some salvias, but butterflies flock to it.

## CULTIVATION

Salvias can be planted all summer long. Just be sure the threat of frost is over before they are planted in the spring. All salvias, whether annual or perennial, want full sun and well-drained soil. The only soil type salvias will not tolerate is heavy soil that stays damp. Fertilizer should be used sparingly, if at all, on the annual salvias and the herb sage. Sage, if grown in ground too rich, will lack flavor and grow leggy. Perennial sages are best given a dose of manure or compost in the fall and lightly fertilized after they finish blooming in the spring. Avoid fertilizers high in nitrogen content; that will reduce bloom and produce rangy plants. A 5-10-5 fertilizer, or its equivalent, is ideal.

In the garden, salvias can go just about anywhere there is full sun.

The herb sage should be planted close to the house to remind you to snip it and use it fresh. It also produces better all summer if kept trimmed back.

Annual salvias mix well with other annual sun lovers, including zinnias, marigolds and petunias. In containers, annual sages add height and look good planted in the center of a trailing type petunia or in the middle of a pot of ivy or sweet potato plant.

Perennial salvias come in enough shapes and colors they could be given a full-sun garden of their own. They also blend well with daylily beds; some gardeners plant them among roses in a bed. A great full-sun combo would be salvia 'May Night' in a bed with the silver-leaved artemisia. Another breathtaking combination is the spiky purple blooms of salvia mixed with the yellow pompom blooms of 'Inca' variety marigolds. Almost makes you high—no Salvia divinorum needed.

## FORGET THE PARSLEY

Remember the Simon and Garfunkel song "Scarborough Fair," which featured the lines "Parsley, sage, rosemary and thyme?" If you want to "grow that song," know that sage, rosemary and thyme are relatively easy herbs to grow. Parsley, on the other hand, is not so accommodating. A relative of the carrot, parsley is slow to germinate and appreciates cool, evenly moist conditions. Good luck with that in the middle of a Kentucky summer. But "sage, rosemary and thyme" just doesn't work, does it?

# Take a walk in the Wiche garden

Fred Wiche would be proud.

His daughter, Jeneen, and her husband, landscape designer Andy Smart, have taken over the 18 acres in western Shelby County that Fred called Swallow Rail Farm—for the barn swallows that flitted through the skies and for the nearby railroad track.

Together Jeneen and Andy have kept up Fred's plantings and gardens, including the row of crabapples leading to the house; have added to and put their own signatures on the perennial beds; and, in a few cases, have undone some of what Fred did. Jeneen and Andy, for example, are planting native perennials along a dam that holds back a small lake on the property and replacing the crown vetch, once considered great for reducing soil erosion, now considered an invasive species.

"Daddy put that there," Jeneen said in an almost conspiratorial whisper.

While Jeneen and Andy are making some changes in the landscape and gardens at Swallow Rail Farm, Jeneen is following directly in her late father's footsteps and is on her way to being almost as recognizable and sought-after a gardening guru as he was. She continued Fred's weekly garden columns in about 20 community newspapers throughout the state and a few in southern Indiana. She hosted a weekly radio show, "Homegrown," with former *Courier-Journal* columnist Bob Hill. She is in constant demand for appearances at garden shows and garden clubs throughout the region.

Where does the inspiration come from for 52 garden columns and radio programs a year?

"It's all right here," Wiche said, pointing to the lawns, trees and flower beds. "This is what I do."

## THE WICHE GARDEN

Wiche's garden is a showpiece of landscaping that fits both the land and the house. It is at once simple and complex, natural yet showy, soothing and startling. The approximately 12 perennial beds ("I lost count," Jeneen said) flow away from the house in pleasing lines that lack the rigidity of some designs; they appear to have been put there by nature, not by human hand. Colors are pleasing and not overpowering, the way Mother Nature plants her garden. But just when the visitor thinks he has the lay of the land figured out, a combination of flowers—purple vervain and coreopsis 'Crème brule,' for example—jumps out and grabs him by the pupils.

Fred bequeathed to Jeneen the "bones" of the garden. He, in consultation with the late Dan Gardiner of Boone-Gardiner Nursery in Louisville, designed and cut the path. The two also added the woody elements, including false cypress, hemlock, Bosnian pines and a tree gardeners should get to know better—Heptacodium, also called "Seven sons tree."

But while Fred was an avid plantsman, he was not much on perennial bed design, Jeneen said.

"Daddy put things wherever there was a place for them," she said.

When she and Andy took over the house and grounds from Fred's widow, Virginia, Jeneen set about revisiting the beds. In one bed, filled with almost nothing but daylilies, she added a diverse collection of herbaceous perennials and ornamental grasses that added color, dimension and shape the bed formerly lacked.

To the bed anchored by a false cypress, she added a monster of a yellow rudbekia, blue and black salvia, 'Drumstick' allium, two perennial verbenas, coreopsis, several 'Chicago Land' series echinacea, panicum ornamental grass and sea holly. She kept one of her favorite daylilies, a strapping melon and orange specimen called 'Rocket City.'

The effect is a pleasing contrast among colors, dominated by blues and purples, shapes and textures. It is a bed that soothes the eye from a distance with just enough occasional contrast to bring the visitor in for closer inspection.

"You have to be patient to get the effect you want, but it doesn't take long really," Wiche said.

Wiche said when she designs and plants a perennial bed she does not first have a final look in mind.

"I like to layer stuff in and look at it for awhile," Wiche said. "I like large clumps, to repeat colors or plants in threes. I want balance and texture, but I don't want it too perfect. And it is always in the middle of summer, on the hottest day, when I'm out making changes."

## Home grown

When she is not altering perennial beds, Wiche is working on her latest project: a vegetable garden where once a field of daylilies grew.

Wiche started with some heirloom tomatoes, greens, potatoes and beans. She gathered an eclectic collection of poles for her bean tepees and set them up in a pattern that looks like a Native American wigwam after he sampled the white man's firewater.

"Everybody that comes out makes fun of my teepees," Wiche said.

Eventually she wants to put in a fence around the garden and add other crops, which she will raise organically.

"I don't want to spray something I eat," Wiche said. "And if I have to spray it, I just won't grow it."

Wiche and Smart also raise a large crop of blueberries (most of which she gives away) raspberries, blackberries, apples and Asian pears.

Another project is replacing the back 5 acres of cool-season grasses with native prairie grasses—big bluestem

## SOME WICHE FAVORITES

Put a plant in the right space, but don't waste time on something that is continually bothered by insects and disease, Jeneen Wiche said. Here are a few plants she believes every gardener ought to at least try:

**Knautia** - low-growing, mounding plant that blooms all summer

**Epimedium** - tough and adaptable shade plant that blooms in many colors but is as, or more, interesting for its foliage

**False indigo** - tall, sun-loving perennial. Try the yellow or white forms.

**Salvia guaranitica** - sun-lover and tolerant of dry, difficult sites. Try 'Black and Blue' or 'Argentine Skies.'

**Hardy geranium** - prolific, low growing sun-lover blooming in many colors

**Allium 'Drumsticks' and 'Star of Persia** - 'Drumsticks' is tall and a striking, deep purple; 'Star of Persia' looks like a static fireworks display.

**Hydrangea 'Annabelle'** - "So easy."

**Panicum** - this ornamental grass, native to the prairie, adds height, depth and texture to a planting.

**Euphorbia** - an annual, grows well in shade

**Echinops** - sea holly, a tough sun-lover

and little bluestem. Once the prairie grasses get established, Smart will add wildflowers.

In her advice columns and on the radio, Wiche encourages gardeners to try new and different plants. But she also requires that plants make it with a minimum of coddling.

"I don't have the time to fuss over something or waste space," Wiche said. "I will try anything to see if it works or doesn't, but if a plant is not giving me much return I'll move on and put in something different."

# Soar with the queen of climbers

No garden of any size is complete without flowering vines. Vines add a backdrop for perennials and annuals. Vines provide another dimension to gardening—up. Vines offer flower shapes, forms and bloom times other plants cannot duplicate.

And the queen of flowering vines is clematis. Nothing can compare with the sight of a 'Jackmanii' clematis, covered in royal purple blooms, covering a white fence in June. Nothing dresses up a drab mailbox like a pink 'Nelly Moser' clematis clambering up the post with some yellow snapdragons worshipping at her feet. A 'Henryi' clematis, with blooms 6 inches across, is a must-have for the all-white garden.

More than 200 species of clematis exist in the world; most are deciduous climbing vines, some with extraordinarily large flowers of 5 to 9 inches in diameter.

The clematis that finds its way into most home gardens is the 'Jackmanii' clematis and the clematis hybrids. But some gardeners like the less showy but still spectacular sweet autumn clematis (Clematis maximowicziana) or virgin's bower (Clematis virginiana) for their sweet smell and later bloom time. The former is a rampant vine that perfumes the air in late August; the latter is a good addition to the native plant garden as it is indigenous to the eastern United States. Virgin's bower will also tolerate damp soils.

The 'Jackmanii' clematis most familiar to gardeners is the royal purple-flowered variety that climbs rapidly and blooms profusely in early to mid-June and sporadically thereafter. 'Jackmanii' clematis have flowers 5-7 inches in diameter.

To some, the clematis hybrids are gaudy additions to the garden with their huge flowers and lipstick-loud colors that range from brilliant white to deep color. Many are bi-colors and some double forms have been bred. Many gardeners bring the flowers inside to float in a bowl of water. The fluffy clusters of seeds that follow the blossoms are also widely used to dress up flower arrangements.

As stated above, 'Henryi,' with its huge, flat flowers is, the classic white clematis. 'Comtesse de Bouchaud' and

'Charissima' are popular pinks. 'Ernest Markham' is a classic red, and the red of 'Niobe' is so deep it is almost black.

No clematis is true blue, but 'Lady Betty Balfour' is close, being a deep blue with touches of purple; 'General Sikorski' is lighter, closer to lavender blue than 'Lady Betty.' 'Nelly Moser' is the standard bicolor clematis; her blooms are purplish-pink with a red center bar. 'Belle of Woking' (white) and 'Duchess of Edinburgh' (deep blue) are popular double varieties.

Clematis can be planted through June. Local garden centers may carry only two or three varieties, so this is one plant you may have to hunt for in garden catalogs or on the Internet if you are looking for a clematis beyond the common.

## CARE

Despite their reputation for being finicky, clematis are fairly easy to establish and grow. They need a light soil rich in organic matter for their roots, support for their vines and sun for the best flower show.

Have your support in place before you plant the clematis; the stems are easily damaged and difficult to move once they get growing. A trellis or fence makes a good support; a clematis will not cling well to a smooth surface such as a metal post. Clematis plants are often sold already climbing a small support.

If the soil is not rich, dig a hole and add purchased topsoil or dehydrated cow manure. If your soil is acidic, you may want to add some lime to the planting hole; clematis likes a soil pH of 7 or higher. Do not over-fertilize or you will get lots of vine growth with little flowering.

Clematis are said to "like their feet in the shade and their head in the sun." The easiest way for the gardener to achieve sun and shade simultaneously is to mulch the base of the plant. Some gardeners place rocks over the roots. Another plan is to surround the clematis with shorter-growing plants, such as daylilies, to keep their feet out of the sun. Once established, clematis vines are drought-tolerant, but you may want to water them during dry spells during the first year of planting.

Clematis has no insect enemies to speak of, but gardeners are often dismayed when a clematis that has thrived for years suddenly shrivels and dies. Clematis wilt is a soil-borne disease that strikes the plant suddenly, usually as the plant gets growing vigorously in mid-to late spring. You will first notice shriveling, drying leaves and then whole stems die. You can try cutting out stems as soon as the wilt appears, but most plants are likely to die anyway. If the plant dies, do not plant another clematis in the same place for at least three years. The disease does not appear to spread to other plants.

Some gardeners get squeamish about pruning clematis. Most of the large-flowered hybrids and 'Jackmanii' hybrids bloom on new wood. That means you should cut them nearly to the ground when the plant is dormant. Some of the large-flowered hybrids, such as 'Nelly Moser' bloom on old wood and should be pruned after flowering. If you are in doubt about what kind you have, you can ask your nursery worker or simply watch the plant for a year and see how it blooms. Clematis is such a beauty queen, you are going to spend a lot of time looking at her anyway.

# Take it easy out there

If you are reading this book, chances are gardening for you is more than passing interest. It is something you do for fun and relaxation. Beautifying your home space and bringing in fresh fruits and vegetables are a welcome bonus.

For many, gardening—weeding, mulching, spraying, pruning, planting—is a delicious change from the world of everyday work—the work of typing, driving, selling or assembling. Gardening should put us back into the original paradise and away from making a living by the sweat of our brow.

But often—and it happens by increments—gardening can become more work than pleasure. We bring to the garden all of the ambition, drive and competitive urges we have in the workplace. Our gardens become the corporate ladder we need to climb, the boss we have to please, the fellow workers we have to compete with.

So in our garden we work to establish that exquisite English garden we saw in *Southern Living*. (Never mind that those plants are never going to thrive because they are happiest in England, a whole 'nother climate than ours.) We douse our lawns with chemical cocktails so ours will look as lush and green as the neighbors'. And we compete to have the showiest roses, the earliest tomatoes or the biggest blooms on our peonies. That is not gardening—that is neurosis.

And then when we get tired of all of that stress, we hire a lawn and landscaping company to do the work for us, we buy all of our food from the grocery and we give up the pleasures of gardening altogether. What a shame!

I used to plant a huge vegetable garden, something like 7,500 square feet devoted to everything from asparagus to zucchini. I worked tirelessly all summer to produce bushels of produce that I froze, canned and jellied. I felt like the world would end if come February we had to buy a can of green beans or a bag of frozen corn. We had plenty of food from the garden, true, but we gave most of the produce away. And I spent more time in the vegetable garden than I did on anything else all summer long.

Do not let yourself get on that treadmill. Try thinking of your garden as you think about your children, or favorite pet if you do not have children. A garden, like a child, is a work in progress and never completed. Gardens, like children, need to grow slowly and naturally. Children don't come to us as full-grown adults; gardens take time to grow, a process that should give us pleasure, not stress. Gardens, like children, are good in themselves. Gardening, like parenting, should come naturally or not at all. You can't raise children by a book. Don't let some garden book writer stress you out about your garden.

Several years ago, I just stopped letting the garden worry me. I stopped worrying about growing enough food to feed the family—and 10 other families besides. Now my vegetable garden is less than half its former self but still provides us with most of the fruits and vegetables we use in the kitchen. We occasionally have to buy a bag of corn or peas, but that is not the end of the world.

I also stopped fretting over plants that will not grow for me. I used to religiously plant azaleas and rhododendrons every year because they look so good next to the old houses in downtown Louisville. But despite buying peat moss and soil acidifiers and fertilizers and watering all summer, the plants would expire before the end of the season. Now I grow spirea, hostas, daylilies—plants you cannot kill with a jackhammer—in place of rhododendrons and azaleas.

I would suggest you do a stress inventory in your garden and determine what plants are giving you the heebie jeebies and get rid of them. Perhaps it is hybrid tea roses. Replace them with old roses or shrub roses. How many dogwood trees have you killed trying to keep up with the Joneses across the way? Dogwoods will die if you look at them sideways—try replacing them with redbuds or serviceberries. Does your bluegrass lawn make you crazy? Replace it with fescue. Or learn to enjoy the subtle beauty of dandelions and chickweed.

And another thing. Make a place in your garden to relax, to entertain friends, to read or just to watch the play of sunlight on green leaves. Put out a bench where you will be tempted to sit for a spell while weeding. Put out a table and chairs where you can sit and talk with your spouse or children. Put a picnic table underneath your biggest shade tree, and plant a few fragrant flowers nearby. Then entertain friends and family in the garden.

Make the garden a place where you come to to relax, like the den or the living room. Don't make it a second job.

## EASY REACH

Can't bend over and kneel much anymore? Don't give up gardening. You can build raised beds of concrete blocks 3-feet high filled with dirt. Plants will grow beautifully because of the excellent drainage, and you never have to kneel down or bend too far to weed or pick produce.

# Make room for flying flowers

Call them floating flowers, fluttering jewels or moveable Monets: Butterflies are a beautiful addition to any garden. Invite them all—the monarchs, swallowtails, sulphurs, hairstreaks and viceroys—by planting a few of the flowers, shrubs and trees they love.

Butterfly gardens are growing in popularity as gardeners realize that butterflies bring their own special beauty and magic to the garden, their wonderful colors and patterns enhancing any bloom they visit. The deep purple bloom of a butterfly bush, striking as it is, suddenly becomes a work of art when a black and yellow swallowtail alights for a sip of nectar.

Over the course of the summer, butterflies will come to nearly any garden that has a few blooming plants. Creating a garden that will actually lure them in, sometimes in great clouds, requires a little more effort. But not too much. Happily, the plants that attract butterflies are also nearly carefree as well as attractive in the landscape.

## Do no harm

Before you start a butterfly garden, keep in mind a few requirements.

First, butterflies are the adult phase of caterpillars. Most of the insecticides designed to protect our plants, flowers, fruits and vegetables kill caterpillars, and that includes organic insecticides such as bacillus thurigiensis (Bt). No caterpillars, no butterflies. Ideally, to attract butterflies, do not use insecticides anywhere in your yard; definitely keep them out of the butterfly garden.

Second, butterflies use sunlight to warm their bodies before taking flight. So butterfly gardens should be placed in full sun. Not surprisingly, most of the plants butterflies favor also grow best with at least six hours of sun. Isn't nature grand? Butterflies are also less active on windy days; a garden that offers some shelter from gusts will be appreciated.

Finally, in the natural world butterflies are most common at the edge of habitats, where trees meet shrubs, which meet flowers and grasses. The best butterfly gardens are those that mimic that pattern with a three-layer design of trees, shrubs and flowers, or shrubs, tall flowers and shorter flowers.

Because butterflies are most active in the summer, they are most attracted to plants that bloom from June through mid-September. They are also attracted to plants that produce nectar, preferably those with a flat surface for landing.

Among the trees that will thrive in Kentucky gardens and also attract butterflies are willows, dogwoods and fruit trees, including crabapples. Good shrubs include buddleia, wax privet, viburnum and hydrangea. Popular flowers for butterflies are asclepias (butterfly weed), yarrow, zinnia, marigold, daylily, rudbeckia and lantana. All of these flowers are not only attractive to butterflies, they should be attractive to gardeners, too, because they are low-maintenance. They will thrive in nearly any kind of soil and are bothered by few insects, making it unnecessary to risk harming the butterflies by spraying insecticides. Do, however, keep the flowers dead-headed throughout the summer so they continue to bloom and attract the butterflies throughout the season.

## THE GARDENS

Let's plan some butterfly gardens. If space, time or money are short, a simple butterfly garden can be nothing more than a single buddleia, or butterfly bush. Butterfly bushes bloom from June or early July through frost and are magnets for butterflies. While buddleia were ignored by gardeners for most of the last century, they have regained popularity within the last decade, and dozens of varieties are available. While scientists are not sure how butterflies see color, the insects seem to be most attracted to the colors blue and purple. So while you are seeking a buddleia at your favorite garden center, look for the varieties 'Nanho Purple' or 'Black Knight.'

If you have a little more space, plant two butterfly bushes, perhaps a purple and a pink, and on the downwind side add three or four yellow daylilies. In front of the daylilies, plant some white marigolds. Or another combination: Try two purple butterfly bushes for the background; put three or four 'Coronation Gold' achillea in the midground; mix some white marigolds and orange asclepias in the front of the garden.

Lots of space but not much time to tend a garden? Try this easy-care butterfly garden: Put two butterfly bushes in the background with two wax privet bushes in between. Row two can include a planting of yellow rudbeckia (black-eyed susans) mixed with purple coneflowers. Fill the front row with white zinnias, yellow marigolds or dwarf potentilla. The garden will attract not only the butterflies but the admiration of your summer visitors.

Now, while you're enjoying the flying artwork in your garden, people who plant for butterflies are doing more than just adding another beautiful dimension to their gardens. They are also providing a refuge for the insects.

Butterflies, like our songbirds, are increasingly harassed by the encroachment of people into their habitats, to the point that many species have become endangered or extinct within the last century. Developers are taking away the plants they need to lay their eggs and feed their caterpillars. When butterflies venture into our habitats, they are often killed by the chemical soups we have concocted to keep our foods, lawns and gardens picture perfect.

If you want to delve deeply into the subject of butterfly gardening, check out the book *Butterfly Gardening: Creating Summer Magic in Your Garden* by the Xerces Society and the Smithsonian Institution. The book has stunning pictures of butterflies and some excellent ideas for bringing them into your world.

## NOT JUST FOR CATS

Catnip, also called catmint, isn't just for making cats swoon. Butterflies love it, too. Try the variety called 'Walker's Low.' It will start blooming in mid-to late May and bloom all summer. Makes a great edging plant, and butterflies will thank you.

# Water, water everywhere

Gardens certainly can suffer from a deluge of rain in late spring and early summer.

If, after a rain-drenched spring, you lose plants early in the summer, or have plants unexpectedly wilting and dying, they may have been drowned out in April and May and have been dying a slow death from the suffocation of their roots caused by standing water. Perennials, shrubs, even trees, succumb to too much water just as readily as they succumb to drought. Most plants, in fact, handle drought better than too much water.

If you have places in your yard or garden that do not drain well, where water stands after a heavy rain for a day or more, you have three options. One is expensive and impractical. One is moderately expensive but do-able. The third is the least expensive but requires the most forethought.

Option one is tiling—digging trenches across the yard and putting in drain tiles to whisk away the water. Farmers do it to drain large grain fields, but it's impractical.

Option two is building raised beds over areas that drain poorly. Raised beds, or berms, are popular with garden designers. If you build them high enough—12 inches at least—above ground level they will get your perennial plants' and small shrubs' feet out of the water. Artistically laid out in curves and lines, raised beds and berms can be an asset to the landscape and solve the drainage problem at the same time.

But it takes a lot of topsoil (and filthy lucre) to build even a modest-sized raised bed. Also, raised beds will solve the drainage problem for smaller plants, but for larger shrubs and trees, which send their roots down more than a foot or so, a raised bed does not keep the roots out of water. Finally, raised beds will dry out much more quickly than the surrounding landscape and will need to be watered more frequently. Commercial buildings and some homeowners install irrigation systems to handle the watering, but these are expensive. The best solution, in my opinion, is to let wet spots be wet spots and put plants there that like, or at least tolerate, having their feet wet for hours or days at a time. It's true, that limits your planting options, but that's OK. Consider it a challenge to your creativity.

Let's look at some plants that will tolerate, even thrive, in wet conditions.

## TREES

Many fine trees are happy growing in wet places; some can grow in standing water. One of the best in my book, is the bald cypress. Its Christmas tree shape and striking size make it a standout. Yes, that's the tree that grows in the swamps of Louisiana, but it will grow here and be quite happy in the low spot in the yard. Just remember, the bald cypress can reach a height of up to 100 feet, so plant it away from roof lines and power lines. A smaller evergreen tree that likes wet feet is the sweet bay magnolia. It has all the virtues of Southern magnolia in a tree that likes it wet. Other trees to consider for that wet place in the yard are: possum haw, red maple, river birch and water tupelo.

## SHRUBS

Shrubs that like it wet are fewer in number, and some are too little used. Summersweet, or clethra, is my favorite. It blooms in July and August, smells heavenly and can stand wet soil. 'Hummingbird' is a good cultivar. Another favorite is calycanthus; I like it because my grandmother was fond of it, its inconspicuous blooms smell like strawberries and pineapple, and it will tolerate wet conditions. Other options are spicebush, inkberry, red-osier dogwood and Virginia sweetspire.

## PERENNIALS

Perennials that like to wade are fairly common. The irises—Japanese, Siberian and yellow flag—will not only tolerate wet soil, but they can grow in standing water and often are planted in water gardens. Many native perennials also will tolerate wet spots. Joe pye weed is one. So are cardinal flower and pitcher plants. Other perennials that like it wet are canna, rose mallow, marsh marigold and meadowsweet.

## GRASSES

Switchgrass, rushes and gamma grass can take the occasional long drink and not pass out.

If you have difficult situations, like wet spots or hot, dry places, or just want plants that produce flowers for butterflies, you might want to check out *The Southern Gardener's Book of Lists* by Lois Trigg Chaplin. It has lists of plants for all kinds of places and for all kinds of purposes. It's published by Taylor Publishing Company of Dallas, Texas.

## DOUBLE DIGGING

Those of you with tight clay soil that drains poorly need to find a bored teenager and get him or her to double dig your garden. The process involves digging about 2 feet deep into the soil and replacing the bottom layer with improved soil, manure, etc. The result will be a bed that has excellent drainage. Just understand that it's going to cost you at least an iPhone5s.

# Growing tomatoes

As hard as it is to believe, before the early 1800s Americans refused to eat tomatoes! They considered the fruits, which they called love apples, to be poisonous. Probably the tomato's close relatives, deadly nightshade and henbane, were responsible for the tomato's bad reputation.

Fortunately for BLT and ketchup lovers everywhere, a New Jersey farmer early in the 19th century took a peck of tomatoes to the courthouse, broke out a knife and salt shaker and consumed them right there on the courthouse steps.

When he failed to keel over in front of the crowd that had gathered, the tomato's reputation as an edible fruit began to be secured.

Today, according to a Gallup poll, almost 90 percent of all gardeners grow at least a few tomatoes. Even in the most upscale subdivisions in Louisville and Lexington, places where any hint of agricultural activity is frowned upon, some homeowners put in a few tomato plants. That's because anyone with even half a taste bud knows how superior homegrown tomatoes are to their store-bought counterparts.

I have always considered growing tomatoes embarrassingly easy, but I get more questions from gardeners about tomato-growing techniques than just about every other gardening subject put together.

So here's the way I do it. It is my way and not *the* way, and you ought to feel free to experiment.

### Tomato tricks

Before you buy that first plant, decide where to put your tomatoes. Tomatoes need a minimum of six hours of sun, so the north side of the house is out.

Also watch your yard after a heavy rain. Anywhere the water stands for more than a day is a place you want to avoid putting tomatoes, or almost everything else but bog plants.

Once you have chosen a site and worked the ground with spade or rototiller, you are ready to pick out plants. Most garden centers offer varieties that do reasonably well in your area, but, truthfully, some are much better than others. Some of the varieties that have produced well and have borne good-tasting fruits for me are 'Celebrity,' 'Big Beef,' 'Better Boy,' 'Marglobe,' 'Rutgers' and 'Supersteak.' A good garden center ought to stock at least one of those.

When selecting among plants, look for those with dark green leaves and stems just smaller than a lead pencil. Buying plants in bloom or with small tomatoes already formed will make for early tomatoes and neighborhood bragging rights, but in the long run they are poor producers. Remember, too, that a few well-grown plants will produce a lot of tomatoes, so control yourself at the nursery.

The traditional day to plant tomatoes in Kentucky is Derby day. Most years that is safe, but not always. Watch the weather. If strong cold fronts are predicted it would pay you to wait until the weather moderates. If you get caught and have to cover your plants, always use cardboard, cloth or paper, never plastic or metal.

To plant, use a trowel to make a hole for the plant deep enough to bury it nearly up to its first branching stems. Put in the plant, fill the hole with water and pat the soil around it.

If you have been troubled with cutworms, which chop off plants at soil level, this is the time to foil them. Put some Sevin dust around the plant or use the cardboard roll from toilet paper to encircle the stem. Somehow the worms don't have enough sense to climb to the top and fell the plant from above.

Space tomato plants at least 3 feet apart; 5 or 6 would be better.

If you choose, mulch around the base of the plant. Black plastic will warm the soil and speed up ripening. Organic mulches such as straw will keep the soil cooler, and since tomatoes love heat, you should wait until early June to apply them.

I put my plants in cages, but staked tomatoes or tomatoes left to sprawl also work. However, staking or caging tomatoes means you will lose fewer fruits to rotting.

In reasonably fertile soil tomatoes do not need fertilizer until the fruits are about the size of a quarter. Then you can add half a cup of 5-10-10 fertilizer per plant, a liquid plant food such as Miracle-Gro, or several shovels full of compost. To keep the plants bearing, it's a good idea to fertilize again after the first fruits ripen.

Tomatoes have a few pests, but none serious. In some years tomato hornworms are troublesome. These ferocious-looking green caterpillars are actually quite harmless—at least to you. You know they are there when you see leaves missing and dark green droppings on the plants. Pick off the hornworms and dispose of them.

One last thing, and this is a rule you have to obey. Once you have gathered your crop, don't let them sit in a sunny windowsill to ripen. You are not helping them ripen; you are cooking them.

## BEST TASTING?

Everyone has a different opinion about what makes a tomato taste great. Some like them sweet, some tangy, some juicy, some a little more firm. In the few "polls" I've seen of flavor comparisons, the hybrid Fantastic and the heirloom 'Brandywine' get the most raves. But grow what tastes best to you.

# Vines: gardening up, not out

Gardeners who think of their plant space in terms of length by width are missing an entire dimension: up!

Gardening vertically, with vines and twining plants, has many advantages, not the least of which is that in a small space adding a vine that heads toward the sky may be the only way to crowd in another plant.

Garden designers also use vines and their supports to add boundaries to gardens, especially useful where the garden is large, or of a rambling shape. Climbing roses over an arch, for instance, can mark the entrance from one part of the garden to another.

Vines can also form a cozy nook or create a shady place in an otherwise bright vista. Grapevines climbing over an arbor with a bench underneath, for example, can invite the gardener or visitor to sit a spell and get out of the hot sun.

Vines can create a visual backdrop for other flowers. Flowers that bloom pink, yellow or white look good against the deep green foliage of ivy, autumn clematis or climbing hydrangea. Yellow daylilies such as 'Stella d'Oro' are a sight to behold fronting a backdrop of clematis 'Jackmanii,' with its royal purple blooms.

Finally, vines can dress up what they cling to (like the side of a shed), add a touch of antiquity to a brick wall, or hide undesirable sights (like an old stump or a rusty fence).

For the most part, vines, once established, are rapacious, undemanding growers. Many, in fact, can quickly get out of bounds, and the gardener needs to plan to do some occasional early spring or mid-summer pruning to keep them under control.

Vines, it is said, like to keep their feet in the shade and their heads in the sun. With few exceptions, they bloom poorly except in full sun. The easiest way to keep the roots cool while the vine takes its rays is to keep a layer of mulch at the base of the plant. During the first year of planting, vines should also be given a deep drink of water, especially during dry spells.

Vines are best planted in rich, well-drained, humusy soil in the spring or fall. Fall-planted vines could use a light dose of fertilizer just before their ordinary bloom time. Choose a fertilizer with a high second number, such as 5-10-5.

Because most gardeners have room for just a few vines, the most difficult part of introducing them to the garden may be picking just the right one.

Many will choose climbing roses. Certainly, there is nothing like the sight of a red climbing rose draped over a white picket fence to provide the senses with a heady treat of sight and smell. But be warned that climbing roses can be tricky to prune correctly and difficult to protect from our quirky winter weather.

Another vine on many gardeners' must-have list is the climbing wisteria. It does present the picture of summer's ease with its vigorous vines clambering over a white arbor, heavy with racemes of purple flowers. But the wisteria is notoriously difficult to bring into bloom; some gardeners spend half a lifetime trying. And the gardener who plants wisteria on anything less stout than a telephone pole is asking for trouble. The plant has been known to pull down a house's siding, gutters and awnings.

If you want vines for your garden, you might try these three perennial vines that are easygoing and require little more than an occasional pruning and light fertilizing, if that.

## Clematis

Called the "queen of vines" with good reason, clematis' blooms can grow to the size of your hand and literally cover the vine. Colors range from deep purples and reds to white and pink. Some are striped along the veins. Clematis should be given a trellis or like support; its delicate fingers do not grasp as easily as do some vines' tendrils.

## Trumpet vine

Trumpet vine, a vigorous native, could climb the Empire State Building. Its bright orange, trumpet-shaped flowers make a bold statement in the garden, and its deep green foliage is a good foil for lighter flowers placed in front. The hummingbirds that flock around it are a bonus.

## Climbing hydrangea

Gardeners who have something to hide may let climbing hydrangea do the job. The plant can be slow-growing for the first few years, but once established, its white flowers against deep green foliage are outstanding. It also has the advantage of thriving in shade, a rarity among climbers.

Two annual vines have also become popular among gardeners for summer color and upright growth: the hyacinth bean and the morning glory.

Hyacinth beans are planted in late spring and rapidly grow to 10 feet or more. The flowers are delicate and purple, the beans are purple, and even the stems are purple.

Some gardeners think of morning glories as weeds, and they can be invasive, but like wandering children can learn their place if trained properly. That place can be over a fence, along a rock wall, up a trellis or around a post. Hybridizers have produced morning glories in dozens of colors with flowers larger than the wild ones. 'Heavenly Blue,' is one of the best.

## BAD GUY

Not all garden vines are desirable. One of the nastiest, most invasive, is the euonymous called wintercreeper. It will climb trees, eventually covering every branch and ultimately choking them. It is very difficult to eradicate. Keep it out of your garden. Gardeners, unfortunately, are to blame for letting wintercreeper escape into the wild.

# 10 totally terrific tomatoes

A few years ago, my family and I went to Hawaii. At the restaurants on the Big Island, just a few slices of heirloom tomatoes sold for $6 to $12, expensive even by Hawaiian standards. It occurred to me that if I could fetch those prices, my dozen or so heirloom tomato plants back home had about $100,000 worth of tomatoes growing on them. That would be a better return from the garden than planting you-know-what.

It also goes to show that the public has come to appreciate the flavor of the old-fashioned tomatoes compared with their rock-hard, pick'em-green-and-gas-them-till-they-turn-red, store-bought counterparts. Heirlooms—tomatoes developed before World War II—typically were grown for their taste, not for their looks, their ability to ship long distances or their perfectly round, red appearance. In fact, some of the old heirlooms are downright ugly. But due to the resurgence of eating locally grown produce, the appreciation of fresh foods and natural flavors, heirloom tomatoes are popular, and you can find seeds and plants of heirlooms in nurseries and garden centers. In fact, the array of heirloom varieties available can be bewildering. While you need to experiment to find the ones that best fit your growing conditions and palate, I'll offer a list of 10 varieties that many consider the best of the "oldies."

## THE HEIRLOOMS

- 'Brandywine' — In taste tests, the

Amish heirloom 'Brandywine' almost always ranks as the best-tasting tomato in the world. It has a rich blend of sweet and acid flavors unmatched by modern varieties. 'Brandywine' is popular, widely available and has been produced in red and yellow varieties; most connoisseurs say the original pink 'Brandywine' is the best.

- 'Abraham Lincoln' — Similar to 'Brandywine' in size and flavor, the original 'Abraham Lincoln' is a good selection if you like your tomatoes large and red. It's probably a little easier to grow than 'Brandywine.'

- 'Mortgage Lifter' — Is a fine, old large red tomato with an interesting story attached. The variety was so popular in the pre-World War II era that its breeder, a West Virginia man named Radiator Charlie, paid off his mortgage by selling plants for $1 apiece.

- 'Cherokee Purple' — You have to get past the color—sort of the combination of the black, red and purple of a punch-drunk prizefighter's face—to enjoy one of the best-tasting tomatoes you'll ever eat. It's very productive in Kentucky, too.

- 'Mr. Stripey — Unlike 'Cherokee Purple,' 'Mr. Stripey' is a pretty tomato. It is mostly yellow, streaked with red. Its slices look great on a plate next to all-red tomatoes. If you have problems eating tomatoes with normal or high acidity, you might try the low-acid 'Mr. Stripey.'

- 'Marglobe' — The tomato I consider the workhorse of heirloom varieties, 'Marglobe' looks like our vision of a tomato, round and red, and it's prolific and easy to grow. It doesn't have the distinctive flavors of 'Brandywine' or 'Abraham Lincoln' but a lot of tomato lovers like it for that reason.

- 'Stupice' — If you want bragging rights for earliest tomato, try 'Stupice.' It originated in Czechoslovakia and grows well in cool weather, so it comes in before most varieties. A small tomato, 'Stupice' produces well all season and has fantastic flavor.

- 'Rutgers' — 'Rutgers' is the variety I remember my grandmother growing because she wanted a good canning variety, high in acid and strong in flavor. 'Rutgers' was created in the 1920s but is still popular among those who "put up" quarts of tomatoes and sauce.

- 'Amish Paste' — Is a giant, bullet-shaped tomato bred for canning and processing. But many like the texture of its thick flesh and small seed cavity for fresh eating. It is my wife's favorite.

- 'Riesentraube' — If you like cherry tomatoes, a couple of plants of 'Riesentraube' will keep you in fruits all season long. Its German name translates "great bunch of grapes," and the fruits do hang on the plants in giant clusters.

Plants of these and other heirlooms are widely available in garden centers. You can also buy plants through catalogs. Seed Savers Exchange, www.seedsavers.org, and Totally Tomatoes, www.totallytomato.com, sell some or all of the above heirloom varieties.

## Growing tips

Tomatoes are the most popular

vegetable in America's home gardens, in part because the store-bought varieties are so bland, but also because tomatoes are among the easiest fruits to grow. Modern hybrids have been developed to resist common tomato diseases and grow in almost any climate. The old tomato varieties, however, can be a little more challenging. Try these growing tips if you plant heirlooms:

• Many of the heirlooms—'Brandywine,' 'Abraham Lincoln' and 'Mortgage Lifter,' for example—are huge plants so you need to contain them accordingly. The little teepee-like wire cages won't get the job done. Get concrete reinforcing wire cages or prune them to *heavy* stakes. Letting them sprawl on the ground will cause many fruits to rot.

• To reduce the splitting and cracking common in older varieties, keep the soil uniformly moist. Tomatoes split when the ground gets dry and then suddenly gets soaked with a heavy rain. Mulching beneath the tomatoes with hay, straw or grass clippings reduces splitting because the ground stays more evenly moist.

• Many of the old varieties are also prone to blossom end rot. Keeping the soil uniformly moist helps that problem, too. Also, adding a tablespoon of Epsom salts around each plant helps. The problem tends to be worse earlier in the season and typically corrects itself as the summer gets hotter.

• Do not fertilize heirlooms heavily. If the soil is good to start with, don't fertilize the plants until the developing tomatoes are about the size of a quarter. Use a cup of 5-10-10 or the equivalent per plant. You can fertilize again after you start picking ripe tomatoes.

• Regenerate your heirlooms. If you want your plants to produce until frost, take a sharp knife and cut out a sucker (looks like a miniature plant) growing from the parent plant, stick it in the ground and water it heavily. It will droop for a few days but eventually take root and grow. It will give you tomatoes late in the season, sometimes after its parent has passed on.

• Eat 'em up. Unlike modern varieties, most heirlooms won't sit on the shelf forever because they have softer flesh and thinner skins. So use them quickly. They taste so good you likely won't let them linger anyway.

## 'ANTIQUE' FOR A REASON

While heirloom tomatoes have added another dimension of flavor to backyard gardening, bringing back tastes we didn't know we lost, some heirlooms are just hard to grow. They are prone to disease, bear poorly, and crack, split and rot easily. As has been said, some of those old tomatoes "are forgotten for good reason."

# Invite insects

This time of year gardeners are in the heat of the Battle of the Bugs. It may be a good time to reassess strategy.

The book, *Insects and Gardens: In Pursuit of a Garden Ecology* by Eric Grissell, argues that we ought to reconsider what we do when we labor so mightily to rid our gardens of every chewing, sucking, flying, stinging and otherwise annoying insect we see. Grissell, a gardener and entomologist, thinks we ought to declare a cease-fire in our attempt to wipe out insects. He even goes so far as to suggest we ought to invite insects into our gardens.

Yes, I know what you're thinking, but let's at least consider his arguments, and if you get a chance you ought to pick up his book.

Gardeners used to thinking of insects as the enemy may have a hard time stomaching his several chapters describing the orders of insects, and that part does get a little bit tedious. I don't much care, for example, that there are actually very few true bugs even though most people think "insects" and "bugs" are synonymous. Nor does it interest me that mites actually belong to the spider family and are not insects at all. But the more than 100 photographs of "bugs" taken by Carll Goodpasture enliven this section.

Grissell says while we in America dump more than 6 billion pounds of insecticides on our lawns, fields and gardens every year, we have done little to lessen the loss of food and ornamental plants to insects. In many cases, we have made matters worse, by unwittingly breeding, through natural selection, super-bugs that are immune to just about any dose of poison we throw at them short of a nuclear hit.

Grissell notes the vast majority of insects in the garden are benign. They do no harm whatsoever, even though the chemicals we use kill them off along with the bad guys.

Then, too, many insects are beneficial. Without the pollinators—bees, wasps and flies—we would have very little fruit and vegetable production.

While some insects are doing the chores of pollinating for us, other "good" insects are busy eliminating the harmful insects, that is, if we would just let them go about their work. Those flying wasps we duck and run from, for example, are eliminating hundreds of leaf-chewing caterpillars from our gardens every day. (Most of those wasps, Grissell points out, do not sting people anyway.) Lady beetles will wipe out colonies of aphids if we do not kill the girls with our sprays first.

Grissell says if we hold up on spraying, most harmful insects in the garden will be held in check by their natural enemies and do little harm. And when insects do a little damage, as they inevitably will, he suggests we be more tolerant of minor cosmetic damage and not reach for the spray can at the first sight of a few holes in the hostas or a few ragged edges on the leaves of a prized shrub.

Gardeners, Grissell says, can encourage this insect balance of nature by planting a mixture of plants instead of a monoculture. Many gardeners, for example, will plant a bed dedicated to roses only. That is like putting up a big "Come and Get It" sign to aphids and other pests. Mixing roses in among other plants, especially blooming plants such as baby's breath that attract beneficial wasps, will go a long way toward keeping harmful insects in check.

Grissell, with the help of Goodpasture's photos, also encourages gardeners to see the beauty of the insect world that will inhabit a garden if we hold our chemical fire.

The caterpillars that we destroy would, if we let them, grow up to be the beautiful moths and butterflies: the dainty sulphur, the intricately-patterned swallowtail, the regal monarch, the stunning cecropia moth. The yellow and black garden spider on her web sparkling with dew in the morning, is a sight every bit as striking as a row of unblemished rose blooms.

Many gardeners these days are planting butterfly gardens with perennials and shrubs such as salvia, asclepias and buddleia. But, as Goodpasture points out, we have little hope of attracting the winged jewels to fly among our blooms if we douse their larval stage, the caterpillars, with insecticides.

The intelligent gardener will pick her fight with the insect world carefully, timing the spray to do the most good and the least harm and avoid an all-out massacre that, in the long run, will do more harm than good anyway.

## INSECT MAGNET

Achillea (yarrow) may be one of the best beneficial insect magnets you can put in your garden. Blooming in shades of white, yellow, pink and red, yarrow offers flat-topped flowers insects seem to favor as landing pads. It is a great mid-summer bloomer, too, and it withstands heat and drought. It's especially attractive to insects called lacewings that prey on many bad guys.

# Plant the lilies of summer

The gardener who says he has a bed of lilies these days is likely referring to daylilies, such is the popularity of that hardy perennial. Let me introduce you to another lily, if you don't know her, one that can be a spectacular addition to your garden. Some gardeners even think she is a little too spectacular.

Like daylilies, true lilies, also called summer lilies, are spring- and summer-blooming perennials that are easy to grow, and add color in the garden during the summer when the spring bloomers have stopped and before the fall bloomers have started.

Like daylilies, summer lilies come in a wide range of colors. But unlike daylilies, summer lilies take color to another level. Their reds aren't just red, they are scarlet; the yellows are startling; the oranges are show stoppers; the pinks are electric; and the whites are so pure they don't seem natural. Many are freckled; some sport blazes. Most have trumpet- or bowl-shaped blooms that are huge—up to a foot across—on plants that can range from 2 to 7 feet in height.

Daylilies in the garden are the pretty girl next door. Summer lilies are the women in the Victoria's Secret ads—so stunning they don't seem real.

That's ironic because the lily that has been in cultivation the longest, the white Madonna lily, was kept by monks in medieval gardens. A symbol of motherhood, its white, trumpet-shaped blooms represented the purity of the Virgin Mary. It was the Easter lily of the past. A medieval writer said the blooms of the Madonna lily were able to show if young women were still chaste, but, mercifully, the details of how that worked have been lost in the mists of time.

Before the 1800s, the white Madonna was about the only lily in cultivation. But trading ships to China and Japan early in the 19th century brought back to England and America summer lilies that became the basis for hundreds of hybrids in the range of colors available today. In fact, white hybrids have pushed Madonna aside to become what we call Easter lilies today.

Modern lily hybrids can be divided into three major groups. Choosing one of each type will ensure blooms from late spring until early fall.

The most common, the Asiatics, grow 2 to 4 feet tall with bowl or cup-shaped flowers that are 3 to 5 inches in diameter. Asiatic lilies are usually, but not always, spotted. 'Marlene' is a brilliant yellow. 'Aphrodite' is a shocking pink with white veins.

Aurelian hybrids have huge, trumpet-shaped flowers on stems that grow 4 to 7 feet tall. Flowers bloom in large clusters in every color except bright red. 'Golden Splendor' is a brilliant yellow with a darker reverse.

Oriental lilies bloom in white, pink or red. Blooms are up to a foot across and are almost flat. 'Casa Blanca' is a pure white. 'Stargazer' is pink with white trim and bold spots. It is the lily that blankets the winner of the Kentucky Oaks on the Friday before Derby Day.

While the three hybrid groups are the most popular, don't overlook the species lilies, including the tiger lilies with their spotted blooms and recurved petals, or the regale lily with its enormous trumpets.

Lilies for the garden can be bought in the spring as single plants in pots. But that can be an expensive way to start a lily bed. Better selection and prices can be had by buying bulbs that will be in catalogs and garden centers in late summer and early fall. Summer lily bulbs are best planted as soon as they are in hand, because unlike tulips and daffodils, they dry out easily. Bulbs should be planted 3 to 4 inches deep.

Site selection is more critical for summer lilies than it is for daylilies both for the plants' happiness and to achieve the best visual effect.

First, go for full sun and well-drained soil. Those are must-haves. Because summer lily blooms are so large and arranged on a single stem, try to arrange the summer lily bed so that it is out of the path of strong winds to keep them from blowing over. The sheltered side of a board fence is one possibility.

Some gardeners like beds of only lilies. They are certainly stunning when in bloom. However, after the blooms are gone, the foliage starts to turn yellow and the lily bed can look pretty ragged. And you can't just cut down the foliage at that point because the stems are still feeding the lily bulb to make that spectacular bloom happen again next year. Don't take out lily foliage until it is completely dried.

The solution is to plant lilies behind or among lower-growing annuals or perennials whose flowers and foliage will hide the yellowing lily stalks. The lilies' "underwear" will be hidden and they will still show to good effect.

It is a good idea to lightly fertilize lilies after the foliage has dried with a high phosphorous fertilizer (the second number on a fertilizer bag) and again just after the lilies emerge next year.

Start looking for the bulbs of summer lilies in garden centers. Plant them in early fall, and they'll charm you next summer. Lily's a lady worth waiting for.

## JUST PUT THEM BACK

One of the problems with having bulbs in the garden, whether they are spring or summer bloomers, is that it's easy to forget where they are and accidentally dig them up while putting in other plants. If you accidentally uproot a spring bulb or summer lily, just put it back. Chances are it won't pout too much about the interruption in its nap.

# Take the melon challenge

Every gardener worth his trowel wants to grow something bigger, better, prettier or more challenging than his fellows.

Some want the first tomato. Some need to grow the perfect rose. Some try for the coffee table-book-pretty perennial bed.

What gets me digging is the challenge of producing watermelons and cantaloupes.

Not only are the watermelon and cantaloupe my favorite fruits—I could eat them every day—it is almost impossible to find good ones in a grocery store, especially good cantaloupes. I always say no matter how much money you have, there are still two things you'll never get if you don't grow them yourself: a good cantaloupe and a good ear of sweet corn.

But melons are a challenge. Part of the challenge is that melons need the right kind of soil, which you can control somewhat; and they need the right weather, which you can't control at all.

Still, if you have ever had a cantaloupe or watermelon fresh from the garden, you know they are worth the try. On top of unmatched flavor, homegrown melons will retain all of the vitamins and minerals they have in abundance because they can be picked at the peak of perfection.

## SPACE, SOIL, HEAT AND WATER

Cantaloupes and watermelons are more exacting in their cultural needs than, say, beans, tomatoes or corn.

In the first place, melons need a lot of room. A watermelon plant will easily grow 6 feet in every direction from home base; a cantaloupe, a little less.

Cantaloupes can be trained to grow up a fence, but the fruit will need to be supported (Fred Wiche recommended used pantyhose) once it gets about half grown. There are bush types of watermelons that take up about half the space of ordinary watermelons, but I have found them to be unsatisfactory producers. But remember that one or two plants will produce considerable fruit if grown well.

Heavy, poorly drained soil is the last place you want your melons to put down roots. Melons need light, almost sandy, soil. You can amend the soil by growing melons in raised beds or digging a large hole and adding about half a bushel of manure or loose soil and planting the melons on top. The manure will add fertilizer to the growing melons and improve the tilth, or texture, of the soil.

Watermelons are native to Africa, cantaloupes to the Middle East, so they like it hot. Melons just do not perform well in a cool, wet summer.

Fortunately, most years it is plenty warm across the state to grow melons. Just don't get in a hurry to plant them. Don't even think about planting melons until mid-May unless you live in extreme western Kentucky. The seed will not germinate in cool soil and will rot. Wait at least until mid-May.

The best trick for adding heat is to buy melon plants or start your own in peat pots and plant them into black plastic. The black plastic will warm the roots and keep weeds at bay at the same time. Buy the type of plastic that is woven, so that it admits air and water but keeps out weeds. Cut a slit in the plastic and set the plant in the slit, being careful to cover the peat pot with soil.

At the same time you plant, be sure to add a good drink of water. Melons are heavy drinkers while they are growing up until about the time the fruits get about half grown. At that time, do not add any more water than what Mother Nature brings. Melons ripened in drier conditions are sweeter than those grown where the soil stays soggy.

While you're watering, add a dose of soluble fertilizer such as Miracle Gro, unless you manured the ground heavily. Melons are heavy feeders. In the old days, melons were often grown in a vacant hog lot to take advantage of the nutrients in the manure.

### Plink or plunk?

How do you know when a melon is ripe?

Cantaloupes are ripe when the fruit will slip away from the vine with just a touch of the finger. Sometimes you will find the fruits already detached.

The slip treatment will not work with honeydew melons. They have to be tugged or cut when ripe. The best way to tell if a honeydew is ripe is to compare it with others in the patch. A honeydew is ripe when its skin turns a duller shade, and sometimes gets a touch of yellow.

You can thump your fingers off listening for a plink or plunk to tell when a watermelon is ripe and still get it wrong. There are three signs of a ripe watermelon. The first is that the little pigtail nearest the watermelon's stem will turn brown. Second, the melon will lose its surface sheen and become a dull green. Third, if you turn it over, the belly will change from white to a cream or yellow. When you see all three signs, break out the salt.

## Big melons?

I am happy to get a ripe melon, but if gardening macho is your thing and you want to grow a get-your-picture-in-the-paper watermelon, try the 'Carolina Cross' variety. It grows to 100 pounds-plus.

One of the biggest cantaloupes is an heirloom called 'Old Time Tennessee.' To give you an idea how big it is, my father-in-law saw one growing in my garden and thought it was a pumpkin.

## Cantaloupe or muskmelon?

Well, technically, what we call cantaloupes are really muskmelons, but if that worries you, you are probably the type who wonders whether a tomato is really a fruit or a vegetable. Just shut up and eat!

## Melon mystery

For years, seed savers looked for a watermelon they heard about that had yellow spots on the melon as well as the foliage. Finally, they found growing in an old Missouri gardener's patch the now-famous 'Moon and Stars' watermelon. It had come within a few years of becoming extinct. 'Moon and Stars' has small yellow "stars" and a larger "moon" on a dark green background. Tastes great, too. It is available through Baker Creek Heirloom Seed Company.

## POOP PILE

In the summer of 2011, I grew an excellent crop of melons, and I attribute that in part to weather and in part to a technique I tried. I raise chickens and rabbits, and when I cleaned out the barn before the 2011 growing season, I piled the manure in a row where I would plant the cantaloupes and watermelons. The manure had plenty of time to rot down before I planted seeds. The vines grew tremendously rapidly and produced dozens of melons in a very short space. The trick is to get the manure down in advance. The melons don't seem to like their manure too fresh.

# Fall

# Berry nice

When we shop for trees and shrubs we usually look for size, shape, flowers and/or fall color. All good criteria for making selections.

Let me suggest that when shopping for plants in the fall (and fall is the best time to plant trees and shrubs) look for another dimension: fruits.

We'll expand the definition of fruits here to include cones, seedpods and berries. Trees and shrubs with colorful fruits or berries, visible cones or decorative seedpods provide interest in winter, after leaves and flowers are long gone.

Consider a few examples:

Sumacs are large shrubs/small trees that positively glow when clad in their brilliant red leaves in October. But the show doesn't end when the leaves fall. Sumacs produce a conical-shaped seedhead that, in the right light, makes the plant appear to be covered in flaming red candles. Another tree that has brilliant red seed pods that stand up like candles is the Southern magnolia.

Beautyberry is a shrub prized exclusively for its berries. In the summer, beautyberry is an ordinary looking thing. But as summer progresses the plant blooms, and then lavender berries line the stems and cling until very cold weather. A striking sight in a dull landscape.

Probably the single best shrub for

berry production, and year-round beauty for that matter, is the viburnum. Viburnums have been bred to produce heavy fruit set. The tea viburnums produce berries that survive a hard freeze. Viburnum berries come in colors from orange to red to blue and black. 'Blueberry Muffin' is considered one of the best blues. When shopping for viburnum, keep in mind that many species and varieties do not bear fruit.

Crabapples may be the best choice for trees with colorful fruits. They have been bred to produce prolific fruits that cling in winter. Crabs have red, yellow, orange, even white fruit. A crabapple branch hanging with red apples coated in snow or ice is as pretty a sight in winter as a cloud of pink blooms is in spring.

Another small tree prized for its winter show is the deciduous holly, frequently called possum haw in the South. The "haws" have a profusion of berries in colors from light orange through deep maroon. 'Warren's Red' deciduous holly may be one of the best for winter show. 'Byer's Golden' is a popular yellow variety. Unlike crabapples, the fruits of deciduous hollies will not turn black after a hard freeze.

## GOOD MATES

A few things to consider when choosing plants for their fruits, berries, seeds and cones.

Many plants, trees especially, will have to reach some age before they produce fruit. Very small crabapples may bloom within a year or two of planting, while the larger species may need several years of growth before they set fruit. And they may need several more years before they set heavy fruit. Some cone-bearing pines not only need to reach a certain age, they need to experience a certain amount of stress before they grow cones.

Some berry and fruit producers need a mate before they will bear. The female deciduous hollies are the berry producers, but they need a boy or two around for pollination. What else is a guy good for? Good nurseries should sell male "escorts" with names like 'Red Escort' and 'Jersey Knight' along with the girls. You usually need only one male for several females. Crabapples do not come in boy and girl, but they produce more fruit if at least two varieties, or even an edible apple tree, are planted in the same vicinity. Choose varieties that have approximately the same bloom time so

### SOME TREES THAT PRODUCE BERRIES

Possum haw, winterberry, American holly, flowering crabapples, sumacs, downy serviceberry, Foster holly, yaupon, weeping yaupon, cornelian cherry, kousa dogwood, hawthorns and Korean mountain ash.

### SOME TREES THAT PRODUCE CONES OR SEEDHEADS

Magnolia, deodor cedar and amur maple

### SHRUBS WITH ORNAMENTAL FRUIT OR BERRIES

Japanese barberry, beautyberry, nandina, Formosa firethorn, tea viburnum, chokecherry, clusterberry, Allegheny viburnum, cranberry viburnum and cranberry cotoneaster.

they pollinate each other effectively.

The good news is that trees and shrubs with berries frequently attract birds. The bad news is that trees and shrubs with berries frequently attract birds. I took a picture last winter of a flock of robins gorging themselves on nandina berries. The sight attracted a gaggle of neighbors who watched the welcome show from the street, windows and porches. But lots of birds, in the wrong place, can be a nuisance. A flock of hungry starlings going after a tree full of crabapples in branches near windows or the family sedan is not so welcome. Moral: Put the berry producers where the birds don't make a nuisance of themselves.

A flock of hungry starlings going after a tree full of crabapples in branches near windows or the family sedan is not so welcome. Moral: Put the berry producers where the birds don't make a nuisance of themselves.

## KEEP OUT THOSE UNWANTED STAINS

Many cultivars of crabapples have been bred to have fruit that is decorative (and edible, so I'm told), especially in the winter. Birds like them, too. But be careful of planting fruiting crabapples and other fruit-bearing trees close to sidewalks, driveways or parked cars. Squashed fruit can leave bad stains.

# Time to paint the spring garden

You can take anti-depressants, but if you're a gardener you can beat the summer's-gone-winter's-coming blues by spending the Indian summer days of October preparing for spring.

You might have noticed all of the little mesh bags of spring bulbs showing up in garden centers, hardware stores, even in the aisles at the grocery. Think that might mean it's time to plant tulips, daffodils, crocus and the like?

You got it. Nothing says springtime in Kentucky like a bed of tulips or a spray of daffodils. But if you want that springtime color, you need to plant in the fall. Some wait until nearly Christmas or later, but doing so means sacrificing quality of bloom or getting no blooms at all.

Here is some advice to help you have the best spring ever:

## Buy quality

When it comes to spring bulbs, size matters. Small bulbs produce small blooms. The little bit more it costs for larger bulbs is readily repaid next March and April.

Because they are primarily grown in Europe, bulbs are measured in centimeters, which is Greek math to most Americans. Just remember that tulip bulbs should be at least 12 centimeters in diameter. Daffodils and hyacinths should be at least 16 centimeters. Smaller bulbs, like crocus and grape hyacinths, should be described as "topsize."

## Choose the right spot

Bulbs are tough, but they are not cast iron. They need good drainage, full sun and reasonably good soil, in that order.

If bulbs sit in wet places over winter, they rot. Those that do not rot will not bloom profusely, and they certainly are less likely to repeat bloom next year.

Bulbs should get at least six hours of sun daily, and eight is better. Even after they finish blooming, bulb foliage needs sun to refresh the bulbs for next year.

Finally, bulbs do not need particularly rich soil, but it should be loose. If you plant in heavy clay, amend the bed with topsoil or peat moss. Fertilize lightly with a low nitrogen fertilizer, 5-10-10, or buy a fertilizer meant for bulbs and follow directions.

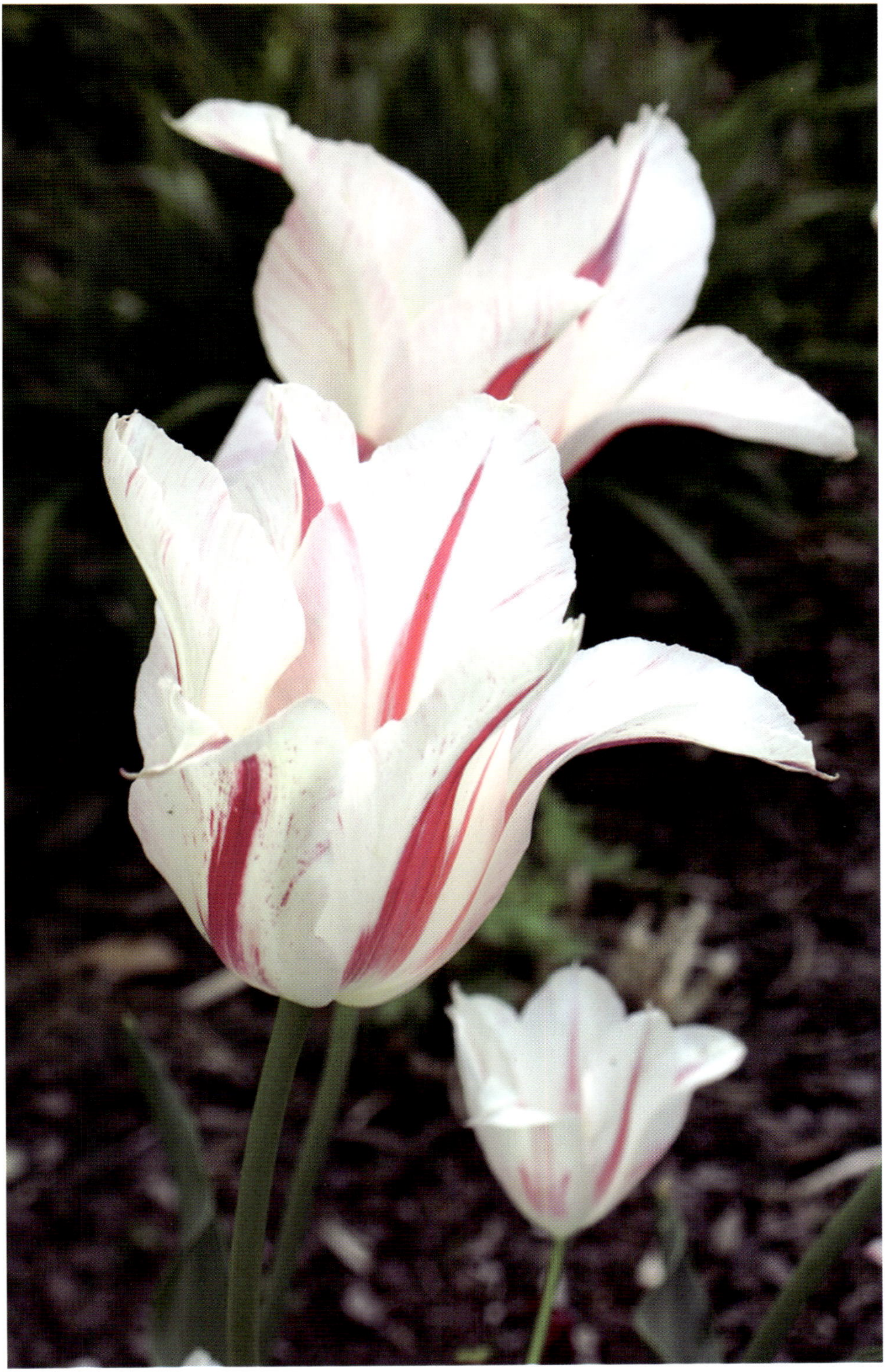

## Plant en masse

A tulip here and there and one over yonder looks like a mistake. A bed or drift of tulips or daffodils looks like a garden. Plant at least 25 tulip bulbs in one spot; plant even more of the smaller bulbs. You can plant fewer daffodils if you like because, given the right conditions, they will spread over the years.

You can make a career out of planting 100 or so tulip bulbs one hole at a time, or you can do it in about half an hour. First, get a shovel or spade and dig out a 5-foot-by-5-foot spot 5 inches deep. Put the bulbs in on a diamond-shaped grid about 4 inches apart (don't line them up like soldiers). Shovel the dirt back over the bulbs and you're finished. If a bulb or two gets knocked sideways, don't worry, it will still grow up.

## Time bloom and coordinate color

If you want to go for the WOW factor in a tulip bed, plant varieties that will bloom at the same time with colors that play off each other. To get the timing right, stick to the same types of hybrids. All Darwin hybrid tulips, for example, will bloom at approximately the same time.

If you are colorblind or fashion-challenged, you may need help with color selections. You might want to check out a bulb company called Colorblends that does the work for you. The company's catalog offers gorgeous, full-color pictures of breathtaking combinations of tulips and daffodils.

## Take care of the foliage

Bulbs bloom, then feed next year's bloom through the foliage. If the foliage is cut down while it is still green, the bulb gets less to eat and will be more stingy with its flower next year, if it blooms at all. Do not cut down the foliage until it has turned yellow or brown. It is also a good idea to give your bulb bed a light fertilizing after the bloom period.

Take care of the foliage and daffodils, hyacinth, crocus, Dutch iris and other bulbs will come back year after year. One way to avoid the ratty effect of browning spring bulb foliage is to plant them among perennials, such as daylilies, that will hide the bulb foliage as the perennials develop their summer growth.

On the other hand, Kentucky gardeners probably need to get over the idea that they can get more than one good blooming year out of tulips. Tulips thrive best where springtime is long and cool, not where spring suddenly becomes summer.

It's probably best to treat tulips as annuals, take out the foliage and plant something in their stead after they finish flowering. If you can't force yourself to do that, plant annuals on top the tulips and hope for the best next year.

### COME AGAIN

Yes, it's true that tulips can be finicky about blooming for more than one year. But the tulips called 'Gregii' are perhaps more stoic than others. I've had a patch of 'Gregii' tulips bloom for the past three years. They're early tulips, usually red, with mottled foliage. Give them a try for a repeat performance.

# Finish that fall garden checklist

Enthusiasm for gardening naturally wanes in late fall. We know frost and cool weather will end most of our outdoor activity soon. And for all but the most hard-boiled gardener, we're getting just a little bit tired of the weeding, picking, spraying and deadheading. Or, as my daughter would say, "We're over it."

That's too bad, because what a gardener does in the fall will go a long way to make next spring—when the long dark hours of winter have made gardening seem an idyllic pasttime again—go so much easier.

Before you put up the shovel and hoe and head for your seat in the easy chair by the fire, make yourself a fall garden checklist of chores to get done in the next couple of months. And on those glorious October days, "Git 'r done."

## Keep producing

If you were new to vegetable gardening in the spring, know that your patch can keep producing well up until the first frost and beyond. Keep picking tomatoes, squash and beans, and they will keep bearing. Some years, we have had tomatoes until nearly Christmas when we have protected the plants from a frost or two.

You can also plant and harvest greens, such as kale or lettuce, in the fall. They will grow slowly but will produce until bitter weather arrives. Fall is also a good time to plant garlic, carrots and spinach to be harvested next spring. Once the seedlings are out of the ground a few inches, and before the first hard freeze, cover with a light blanket of hay or straw and in early spring, you'll have food ready for the table before other gardeners have turned over a shovel of dirt.

## Plant trees and shrubs

All of my readers are required by Walt's Law to plant at least two trees per year. If you didn't plant a tree in the spring, you owe me two in the fall. Fall is the better time of the year to plant trees and shrubs in Kentucky, anyway. Remember to think about the trees that were damaged by ice or wind over the last 12 months, and either avoid those species or make sure they are planted in places where they can't do harm if they fall or limbs come down.

While you're planting a tree, make room for a couple of shrubs, too. Nothing will make a faster impact in your landscape than a single specimen or group of handsome shrubs. An entire shrub border is an effective use of space if your lot size is large enough. Great shrubs for Kentucky gardens are calycanthus, clethra, lilac, viburnum, oakleaf hydrangea, vitex, chokeberry and elderberry. The latter two are relished by birds and are good choices where the gardener wants to stick to native plants.

In the orchard, make note of which trees suffered from diseases in the past, and consider replacing them with disease-free varieties in the spring. In the past, I had a lot of disease in some apples, such as Fuji and Jonagold, but little or no disease on Enterprise or Gold Rush apples. So I decided that Fuji is a great apple to buy in the store, but it just doesn't want to grow here, so out it goes.

## Chop, chop

Cut to the ground the stems of perennials, such as peonies, but leave standing until spring ornamental grasses and seed-bearing plants such as coneflower and rudbeckia. Grasses look better to me in the winter than they do in summer, and the birds will appreciate the gift of seeds when the snow flies. Just be warned that you may have coneflowers, sunflowers, zinnias and other species pop up everywhere next spring, as the birds don't care where they sow. This year, I had zinnias come up among my tomatoes, under a cedar tree and next to the compost pile. I liked the effect, but a more formal gardener may not be amused.

Rose growers usually get too antsy in the fall about putting roses to bed. Wait until we've had at least two good freezes so that the ground is thoroughly cold. Then cut the canes back to about 18 inches and mound up the middle with light soil, compost, even old hay or straw. The goal is not to keep the plant warm but to keep the ground uniformly cold so it doesn't heave from freezing and thawing and so that deep cold doesn't penetrate to the roots.

If you have everbearing raspberries, cut the canes back to the ground, but if you have June-bearers, leave the older canes because they will be next year's producers. Cover your strawberry bed with a light layer of straw after a couple of frosts but before a hard freeze. Other pruning chores, such as reshaping fruit trees, should wait until late winter, though it is always a good idea to remove branches that are dead.

## GET YOUR TOOLS READY

Spring is no time to be doing routine maintenance on power tools. Do that now before you put them to sleep for the winter. Change the oil and sharpen any blades. Some mechanics will say to fill the tank with gasoline and add "fuel saver" chemicals to the tank; others say drain the tank of gasoline for winter storage. I have tried both and feel that the full tank of gas works better, at least for lawn mowers and tractors. I'm not so sure the same is true for small power tools such as string trimmers and leaf blowers.

Hand tools should be cleaned, sharpened and wiped with an oily rag before storage. Some gardeners accomplish this by keeping a bucket of oily sand handy and dipping the tools after each use. Floor wax on wooden handles prolongs their life.

## WRITE IT DOWN

Fall is a good time to make entries in a garden journal. Record what varieties thrived and which failed. Note the dry spells, the heavy rains and wet weather. Write down what plants made it through the summer without showing signs of rust, mildew and other diseases, and which fended off insect pests.

Also, go through the pesticide and insecticide cabinet and note how much product is left over after the season. If you have large quantities left over, you may want to order smaller amounts next spring, or maybe you are buying something you don't use or need.

Finally, write down any observations that will help you improve next year's garden. Remember, next year's garden is always the best one yet.

## DETHATCH?

Many homeowners spend considerable time, effort and money dethatching lawns every year. The idea is to thin out the dead material to allow water and fertilizer to reach the roots better. Everything I've read mentions that dethatching, in most cases, is not beneficial. If the lawn is less than desirable, it's probably better to do a total kill and reseed with a recommended grass such as tall fescue than to dethatch.

# It's September, think spring

It can be a challenge to think about starting a new garden project in September, but if you have ever marveled at the gorgeous displays of spring bulbs in March or April and kicked yourself for not getting the bulbs in the ground last fall when you should have, this is your heads-up.

If you order bulbs from a catalog, do so immediately. If you buy your spring bloomers from a garden center, know they should be arriving by month's end. Optimum time for planting is October.

No spring flowers put on a more stunning visual display in March and April than a bed of tulips. None is more fragrant than a group of hyacinths. The crocus can't be beat for toughness in the face of the vagaries of spring weather.

But for my money, daffodils are the single best all-around springtime flower. Daffodils are nearly as tough as crocus. Only the most bitter cold spring weather will damage them, and then only if they are already blooming.

Unlike tulips, daffodils are truly perennial, coming back year after year in spite of neglect. Old cemeteries and abandoned homesteads still sprout clumps of daffodils long after the man-made structures have fallen to ruins. In fact, daffodils will spread and increase their display every year if left undisturbed. While ground squirrels and voles will eventually decimate a tulip or crocus bed, animals leave daffodils alone—the bulb is poisonous.

With the right selection of varieties, the gardener can have daffodils blooming from late February through early June. No other spring bloomer can match that range. While the miniature daffodils such as 'Minnow' or 'Tete-a-Tete' will bloom in February, the 'Pheasant's Eye' daffodils will bloom up to late May or early June. These latter daffodils are also fragrant.

While it's true daffodils do not sport the range of colors common to tulips or hyacinths, they are no longer available only in white and yellow varieties. Some daffodils have red, salmon, pink or orange trumpets (the cup-shaped inner flower). 'Fortissimo,' for example, has a rich red trumpet against a striking wreath of yellow petals. One of my favorites, 'Salome' has a salmon-pink trumpet against a pure white perianth (the outer ring of petals). Some, like 'Flower Record,' have multi-colored trumpets. A few, like 'Raspberry Ring,' have multi-colored or striped perianths.

Some daffodils are doubles; that is, instead of single trumpets they sport fluffy, many-petaled cups in the center of the flower. 'Flyer' is a good example. Daffodils also come in nodding varieties, the best of which is 'Thalia,' and multi-floral varieties like 'Martinette' that sport yellow and orange flowers on a single stem.

## Cultural requirements

Before we go any further, let's clear up the differences between jonquils, narcissus and daffodils. There aren't any!

They are all forms of narcissus, so you decide what you want to call them and don't let some garden know-it-all try to tell you what you have is one or the other. I like "daffodils," unless I am feeling particularly good about myself; then I might call them "narcissus," named as they were for the Greek mythological figure who pined away and died because he fell in love with his own image.

What you will love about daffodils is their ease of care.

Plant them in the fall, no later than Thanksgiving, though some have thrived and bloomed after being planted as late as Christmas.

Plant daffodils in full sun to a depth of three to four times the height of the bulb. Some people add a bulb fertilizer to the soil at planting. I do not, preferring to fertilize after the spring display.

There are few things you can do wrong, culturally speaking, with daffodils. One is to plant them in a spot with poor drainage. The bulbs will rot. Even so, some varieties, especially the old 'Pheasant Eye' types, will tolerate occasional periods of standing water.

Avoid cutting down the stems after the plant has bloomed. Daffodils, like all spring bulbs, use their stems and leaves after blooming to gather energy to produce another crop of blooms next spring. If you cut down daffodils before they turn yellow and wither, the plants will come up next spring, but they won't bloom. To hide the yellowing foliage, plant daffodils among daylilies or summer annuals.

The other maintenance need daffodils have is an occasional thinning. If your daffodils become crowded and blooms get sparse, you may need to dig up the bed and replant with the largest bulbs 3 to 4 inches apart.

It doesn't take much effort to grow the plant the Chinese call "sacred lily." And they will boost your self-image as a gardener.

## VINTAGE BULBS

Old House Gardens is a neat shopping place for connoisseurs of heirloom tulips, daffodils and other bulbs. The company offers some bulbs that have been in the garden trade since the 1550s. Visit the online store at www.oldhousegardens.com.

# How to pass the soil test

How did your garden grow this spring and summer? If bountifully, congratulations! If poorly, maybe it's because you are doing the gardening equivalent of shooting in the dark. In other words, you're trying to grow plants without getting down and dirty.

You've got to know the soil, brothers and sisters, if you want that garden to grow. Do you know whether your vegetable garden dirt has enough nitrogen to grow a good crop of sweet corn? Do you have enough phosphorous to get fine blooms on your peonies? Is the soil too alkaline to grow azaleas? Don't know? You may need a soil test.

A soil test can give you information about the nutrients the soil already contains as well as its acidity or alkalinity, organic content, mineral content and other vital information. Armed with an accurate soil test, you know what fertilizers in the form of nitrogen, phosphorous and potash (noted as NPK on the bag of fertilizer) and/or amendments to add to the soil this fall and next spring—and what to leave out. Putting fertilizer where it is not needed does three-fold damage: It hits your wallet; it can damage the environment through leaching into the water supply; and it can harm plants. In fact, I believe that many gardeners do more harm to their gardens—and especially their lawns—through over- rather than under-fertilization.

Get a handle on what is really going on under the surface of your garden by getting a soil test.

## WHERE TO GO

Before you start digging, decide where you will have the soil tested. Basically, you have three options: a private soil testing lab, the Cooperative Extension Service and home soil-testing kits.

Private laboratory soil tests are the most expensive of the three options. Expect to pay $40 or more, but often the recommendations are more in-depth than you will get from the Extension Service and likely far more accurate than you'll get through a home test kit.

Kentucky is blessed to have a Cooperative Extension Service that still performs free soil tests for citizens either free or low cost. You will have to take the soil to your local Extension office (you can mail it to private laboratories), and you may have to wait a while for results to come back, especially if you sent the soil sample in during a busy time—early spring and fall.

Garden centers sell home testing kits with which you can mix chemicals and measure everything from nitrogen content to soil acidity. Some are quite inexpensive; those that perform more sophisticated tests can be pricey.

### How to sample soil

Wherever you have your soil sample analyzed, you must do your part to ensure that the dirt you send is an accurate representation of your garden.

First, decide what you are sampling for. Do you want recommendations for a vegetable garden? Flower garden? Lawn? A flower garden, for example, will thrive with less nitrogen and more phosphorous than the typical lawn or vegetable garden.

### Then follow these steps:

- Scrape away any surface litter or plant growth from a small area of soil.

- Using a soil probe or stainless steel tool, dig a hole and collect a slice of soil from the side.

- Repeat the sampling procedure at about 10 different locations on the site.

- Mix the soil cores in a clean plastic or stainless-steel container. You will need about a pint of dry soil. Try not to touch the soil for sampling with your hands or galvanized, brass or soft-steel tools, which might contaminate the results.

- Take the soil to your Extension Service or mail to the lab. Be sure to note your plans for the garden/lawn when you send in the sample or when you consult the chart on the home test kit.

What you get back should tell you what fertilizers you need—and don't need—how much fertilizer to use, whether you need lime to adjust pH or, less likely, another amendment to make the soil more acidic. Depending upon your garden plans, you may need some trace minerals that are missing from your soil, although that is rarely a problem for the general vegetable or flower garden.

After many years of gardening, you may learn to look at the plants themselves and guess what is missing from the soil. But until you achieve garden guru status, take the soil test every two or three years and make sure you know what you're doing out there.

## WHY FIGHT IT?

Most gardeners in Kentucky have either acidic soil or soil that is nearly neutral in pH. But if a soil test shows you have alkaline soil, you may want to consider planting the flowers that thrive in that condition. They include columbine, coneflower, Shasta daisy and globe thistle.

# Live and let live

My attitude toward lawns is different. I like a lawn that is reasonably short and reasonably green, but I am unwilling to work endlessly to achieve front-cover-of-*Southern Living*-perfection.

When we moved we inherited about 5 acres of grass, 3 acres of which are in the front yard, so I put my lawn-care attitude on large display.

The yard has everything in it from clover to dandelions to bluegrass to crabgrass and Creeping Charlie, most the result of the lawn being torn up to build the house and septic system.

I could choose to douse the lawn with a herbicide, apply some heavy-duty fertilizer and then spend my every waking minute during the summer cutting grass.

Instead, I simply mow the lawn with the mower cutting at 2½ inches, let the clippings fall on the lawn and then let nature take its course. Most of the really weedy lawns you see are the result of the grass being cut too short or the lawn having been disturbed with recent digging.

In my experience, the single best thing you can do to eliminate weeds and get a healthy lawn is cut the grass high and often.

When you do that, the taller lawn grasses will eventually shade out most weeds. As you mow and leave the clippings on the grass, you build fertility naturally.

Another way to build fertility naturally is to let the white clover in the lawns grow. Most herbicides are designed to kill broadleaf weeds, including clovers, but clovers have the ability to take nitrogen from the air and feed it to your lawn grasses, all at no cost to you. Eventually, the clovers will do their job so well the grasses will take over their spot and they will move on.

There are two problems with my system of lawn care.

One is that you have to be patient. Mowing the grass high and letting the clippings lay on the lawn will build fertility, but it will take several seasons.

The second is weeds. There will always be a few, probably some dandelions in the spring, some crabgrass or buckthorn in the summer.

But if a few weeds bother you that much, stay up in the tree.

## GRASS CLIPPINGS

The only reason to take grass clippings off the lawn is if you waited too long to cut (shame on you!), and the grass blows out in turf-choking piles. Use the clippings as mulch in the vegetable garden unless you spray with herbicides.

## THE MARK OF FIDO

Seeing brown, circular spots on your lawn? That may be the mark of Fido doing number one. The salts in dog urine cause varying degrees of damage to the lawn, including unsightly brown spots. Watering the spots heavily will undo some of the damage, but you'll probably just have to wait for surrounding grass to fill in. And add lawn training to house training of your next puppy.

# Any way you say it, peonies offer lasting beauty

No matter how much you think you know about plants and gardening, it's a good idea to stay humble. Let me illustrate.

When I was teaching at the University of Louisville, friends seeking gardening advice often accosted me because I was "the plant guy." One day a friend brought in a root piece of a beautiful pink peony she had growing in her yard. I had admired it on a previous occasion and she told me it had once been her aunt's and had been passed down to her.

"Here's that peony I promised," she said. But she didn't say "Pe O ne" as I had always pronounced the plant's name, just as my mother, grandmother and great-grandmother had pronounced it before me. She called it "Pe uh ne."

"You grow a nice peony," I told her, "but you don't pronounce it right." She was equally adamant that I was wrong. Being English teachers, we repaired to the dictionary where, sure enough, her pronunciation was the correct one.

Not only did that incident burst my horticultural hubris, it made me rethink how much my ancestors really knew about plants. But, truth be told, while they may not have pronounced peonies correctly, they could grow them just fine. In May a long row of peonies my great-grandmother had planted sometime in the 1940s bloomed profusely. It's a sight I still associate with the sweetness of spring and the coming of that long luxury of summer vacation for us kids.

However you pronounce "peony," the perennial should be a part of every garden. Peonies fit my definition of perfect plants. They bloom beautifully, are hardy with few disease or insect problems and the plant, even when not in bloom, is handsome. And here's a bonus: The peony is one of the longest-lived perennials you can plant. Plants that were planted in England more than 300 years ago are still alive. The 'Festiva Maxima' peony, developed over 150 years ago, is thought to be the oldest hybrid plant still in commerce. Many 'Festiva Maximas' bloom where the home they stood in front of is long gone.

Peonies come in several species, but the most popular by far is Paeonia lactiflora, which grows from 2 to 3 feet tall with a bushy effect. The plants bloom in all shades of white, red and pink. Hybridizers have added a yellow, but because they are rare, they are expensive. If you have to have yellow, plan to part with $50 or more for a single root.

While peonies bloom in mid-May through mid-June, late September and October are the best times to plant. Some excellent cultivars are 'Festiva Maxima,' white flecked with deep red; 'Sarah Bernhardt,' pink; and 'Kansas,' a deep red.

### CULTIVATION

However you acquire your peonies, take care with planting. Remember, if treated well, they will outlive you.

Peonies generally prefer full sun, but in our hot climate, a little afternoon shade is not a bad thing. Try to find a site for your plants away from prevailing winds. One of the downfalls of the double and bomb-type peony flowers is they become top heavy in rain and wind and flop toward the ground. Some people use wire peony guards to hold up the blooms. Another trick is to plant peonies in groups of threes, and they will help hold each other's blooms off the ground. Make sure the site you choose for peonies drains well.

Peonies like deep, rich soil, so dig a hole as if you are planting a small tree. Enrich the soil with compost or well-rotted manure.

If you have been skimming this column, stop now and pay attention to this: plant the peony root NO MORE THAN 2 INCHES DEEP. Planting too deep is the main reason peonies do not bloom. 2 INCHES. A light layer of mulch to protect the root over winter is not a bad idea but keep that shallow, also.

The peony should bloom within two years after planting. If it does not, and if you haven't planted it too deep, watch to see if the plant gets buds that blacken before they bloom. That is botrytis rot and can be cured by spraying the buds with a spray containing a fungicide or sulfur when the buds break.

Some gardeners feel the need to spray for ants on the peonies. Don't. The ants aren't hurting anything—you or the peony. If you want to bring the flowers in and leave the ants out, just blast with a spray of water. Peonies are the ultimate spring cut flower; their fragrance will fill an entire room. Cut the foliage to the ground after it turns brown in mid to late fall.

After the plant is 2 to 4 years old, you can dig up roots from the outside of the plant and pass them along to friends. Just don't be a smart aleck about it.

## A GATHERING OF PEONIES

Nothing matches the perfume of a vase full of peonies in the house. To gather peony flowers for bouquets, cut them just as the buds begin to open. Leave at least three leaves behind on the stem you cut and don't remove more than half of the blooms from the plant.

# Magnolia, magnificent!

The Northeast has its sugar maples. The Northwest has Douglas fir. The Plains has cottonwoods. The quintessential tree of the South is the Southern magnolia.

Can't you just hear the late Alabama actress Tallulah Bankhead saying the Latin name for Southern magnolia—"magnolia grandiflora, dahlin'."

Like the deep South, magnolias have a timeless charm. Rightfully so. The magnolia family is one of the oldest groups of trees in existence. Dinosaurs probably sipped iced tea and read Faulkner under the shade of magnolias.

Commonly called "bull bay," Southern magnolia, with its large, shiny, evergreen leaves and heavily scented white flowers, stands out in any landscape. Gardeners who plant Southern magnolia should give it plenty of room; it can grow to 100 feet or more.

Kentucky gardeners also need to know that Southern magnolia is near the northern limits of its range in Kentucky. Our winters frequently kill its flower buds. That's not so bad because the tree, even when not in flower, is ornamental. But occasionally, we get a winter that damages the tree itself.

'Little Gem' is one variety of Southern magnolia that is not only more cold hardy than most, it is also smaller than the typical Southern magnolia, rarely getting taller than 40 feet. So it may fit in the typical landscape better than most varieties. 'Edith Bogue' is another cold-hardy Southern magnolia. If you don't find those in your nursery, look for varieties with brown-backed leaves; they tend to be hardier in our climate than those with green or silver-backed leaves.

Southern magnolia is the tree most people think of when they think of "magnolia." But magnolias come in all sizes and shapes, from small shrubs to small trees to huge shade trees, like the cucumber magnolia that are grown more for their shade and distinctive leaves than for their flowers. Fortunately, landscapers are making better use of the other magnolia species in their plantings, and nurseries are stocking more species and varieties. While the Southern magnolia is evergreen, most species are deciduous.

One that ought to be used more is magnolia Virginiana, called sweet bay. Sweet bay looks much like Southern magnolia, with glossy leaves and large white flowers. But it is a smaller tree and is usually deciduous in our climate. Unlike the Southern magnolia, it will tolerate considerable shade and wet soil.

Magnolia soulangeana, sometimes erroneously called "tulip tree," is a deciduous magnolia that covers itself in saucer-shaped, purplish flowers in spring before the leaves come. It is commonly and correctly called "saucer magnolia." Saucer magnolias are usually small, multi-stemmed trees, though some specimens grow as a single-trunk tree and get up to 40 feet or more. Many saucer magnolias bloom early, and their flowers get hit by frosts. That gives the tree the dismal effect of a wedding cake that set out two weeks before the reception. One variety that blooms a little later and usually escapes frost is called 'Alexandrina.'

An increasingly popular magnolia is magnolia stellata, called "star magnolia" for its star-shaped flowers. Star magnolia grows as a small to mid-size shrub and is often used as a foundation plant. Its white flowers in mid-spring look great against a brick or dark wood background. 'Royal Star' is a popular variety.

There are dozens of other magnolias that are hybrids among the various species. Most have white, pink or purple flowers, but some have yellow flowers. When looking for magnolia, check for hardiness (at least zone 6; zone 5 is even better), flower color and size. Magnolias with girls' names — 'Jane,' 'Elizabeth' and 'Ann'—do well in Kentucky. If you buy locally, you are likely to get a magnolia that is hardy in your area.

The Southern magnolia is best planted in spring, but the deciduous magnolias can be planted in the fall; late October is a good time. Before planting, check the ultimate size of the tree and keep the taller varieties away from rooftops, windows and power lines. Magnolias normally grow fairly quickly.

Except for sweet bay, magnolias should be planted in well-drained soil that is on the acidic side. Many gardeners add peat moss to the hole before putting in the plant both to loosen the soil and supply an acidifier. Keep magnolias evenly watered for the first year. Once roots are established, magnolias can withstand considerable hot, dry weather, though they should be watered during a prolonged drought. Magnolias have few, if any, insect pests, so spraying should be unnecessary.

## PRETTY GIRL

Breeders have produced dozens of magnolia cultivars that are crosses between species. One of the prettiest is 'Elizabeth,' a cross between magnolia acuminata and magnolia denudata. A small tree, it covers itself in soft yellow blossoms once it reaches maturity. Plant it in front of dark evergreens for a marvelous show.

# Reflections on a dismal season

Sometimes Kentucky gardeners get hit with a double whammy. The spring is cold, and many plants and blooms succumb to frost. By the time warm weather sets in, drought follows on its heels. I have found that if the season turns dry in July or August not much harm is done. But if the spigot is turned off in May and June, look out!

Add an unusually buggy year, throw in some plant disease and you could have one of the worst gardening seasons in memory. One particularly bad year, my vegetable garden did not produce enough to keep a robin alive. I was able to keep the annuals alive by watering, but as soon as I turned my back to take a vacation, down they went. Hostas scorched. Young trees dropped leaves and turned up their roots. Shrubs wilted and dropped their blooms. Roses refused to bloom after June. And Japanese beetles, cucumber beetles and squash bugs ate what had the nerve to hang on.

Here are some lessons I have learned and some strategies you might try to prevent your garden from looking a little less like a nuclear holocaust the next time we have a dismal season.

## Lesson one: Mulch

One thing I noticed during a rough year is that while the front yard turned as brown as a potato, a strip of grass near the blacktop driveway always grew lush and green. Why? Well, I wondered, too, until I noticed that in my garden the plants that weathered the heat and drought well were those nearest some form of moisture-retaining mulch. Beneath the blacktop driveway the ground retained some moisture so the grass alongside had a spot to dip its toes. In beds where black plastic or porous mulching fabric lay, the plants stayed greener.

Porous mulches, those with tiny holes to let in moisture and air, work best at keeping weeds out and moisture in. Before planting next year, put them down, cut a hole for your plant and cover lightly with a shredded bark or mulching material of your choice.

## Lesson two: Plant trees and shrubs in the fall

Spring is an exciting time to visit your nursery and garden center, pick out trees and shrubs, and take them home to plant. But I noticed that in a trouble-prone year the trees that lived had been planted the previous fall and had time to develop a better root system before dry weather came along. Trees planted in the spring barely developed leaves before the heat and dry weather sucked the life out of them. Trees and shrubs are a major investment, so if you have to plant them in the spring, make sure you have a way to give them about an inch of water per week if Mother Nature turns stingy with the fluids.

## Lesson three: Take advantage of shade

Plants growing in the shade, especially if they can escape the afternoon sun, are more likely to weather heat and drought than those taking a full blast of old Sol. Yes, I know that the plant catalogs and garden books tell you that most plants will bloom best in full sun. And that may be true in a cooler summer or if you live at the latitude of Dayton, Ohio, or farther north. But some shade, especially afternoon shade, is an advantage to gardeners here in the upper South. One of the prettiest borders I have seen during a hot, dry summer was a combination of Asiatic lilies and hydrangeas that were shielded from the afternoon sun by an adjacent building.

## Lesson four: Have a first-overboard strategy

Many communities ration water in dry conditions, and nearly every gardener has to ration his or her time. If drought turns severe and you can keep only some of your plants alive, it makes sense to water the most valuable first and the least valuable last. Of course, if you have an heirloom rose Aunt Edna gave you, better keep that one watered lest you get washed out of the will. Otherwise, the general rule of thumb is to water small trees first, shrubs second, perennials next and annuals last.

## Lesson five: Water deeply

Gardeners spraying a hose over their flowers may be having a jolly time and feel better about themselves when they are finished, but they are not doing much for their plants' survival. It is far better to water once every two weeks to a depth of 6 inches into the soil than to spray only enough water to wet the surface twice a day. When you water lightly, you encourage plants' roots to grow near the surface. Skip a day or two for a vacation and down your babies go. Use a soaker hose or similar apparatus to put lots of water near the roots of plants. If you have a sprinkling type device, such as a Rainbird waterer, turn the water off a few hours before dark. Wet plants on warm nights encourage disease.

## Lesson six: Try the Darwinian approach

Those plants that make it through a difficult year are your troopers. Plant more of them next year. Those that bite the dust can be thrown on the ash heap of your garden's history.

## Lesson seven:

Next year's garden will be better. It always is. At least until next year.

# Make room for big trees

Landscapers, garden designers and garden writers have done a good job of getting homeowners to use smaller trees for small urban and suburban lots. Maybe too good a job.

It is true that large trees overwhelm a small city lot. Large trees make the typical suburban house look like a doll's house. The roots of large trees can buckle sidewalks. A large tree can climb into power lines, take down gutters and damage roofs when planted too close to the house.

But big trees have their place where owners have a little bit more land, and the big guys need to be planted more frequently. After all, Kentucky was once blanketed by a dense forest of big trees—oak, ash, hickory, maple and tulip poplar. We need big trees for their shade, for their beauty, for their ability to feed and harbor wildlife, and for the good of the planet's oxygen exchange system.

We also need them for our souls. Redbuds are charming. Dogwoods are cheering. But a 200-year-old red oak, reaching 80 feet or more toward the crystal-blue sky of autumn, its leaves shellacked in red, is awesome in the not-overused sense of that word.

Gardeners planning their fall nursery trip should think about including a big tree for their landscape along with the perennials and shrubs they buy. And if you don't have room for one yourself, give one to a friend or neighbor with more yard space and have a tree planting party. With luck, you'll still be friends and neighbors when the tree is casting large shade.

Before we talk about some big trees you might want to consider, let's go over the rules: Know the ultimate height and width of the tree. Do not plant big trees near or under power lines. Do not plant the tree where it will grow over a roof or into the side of the house. Have some consideration for your neighbor here, too.

Second, consider cultural requirements. Most large trees require excellent drainage; do not plant them in low spots where water stands for more than half a day. Large trees require more or less full sun. Large trees may require watering their first year as they get established.

If you are going to plant the tree yourself, consider buying smaller specimens from the nursery. Trees more than 8 feet tall ought to be planted by professionals. And remember the planting rule: Dig a $10 hole for a $5 tree. In other words, make sure the hole is large enough to accommodate the root system. A large tree is going to be a long-term investment and the focal point of your lawn or garden; take time choosing and planting it.

Dozens of large trees are native to Kentucky, and many are good landscape prospects. Here are 10 of the best:

• Ginkgo (Ginkgo biloba), the tree that has been on Earth long enough to have met tyrannosaurus rex. Ginkgos have fan-shaped leaves and buttery yellow fall color. Kew, or Magyar, are upright male varieties you can plant nearer to structures than average for the species. Always buy a male ginkgo; females have fruit some say smells like dead animals.

• Red maple (Acer rubra), one of the fastest-growing of the large trees of the forest and one of the most beautiful in fall. Cultivars 'October Glory' and 'Red Sunset' are smaller than average for the species and show brilliant red fall color. 'Scarlet Sentinel' is an upright variety.

• Kentucky coffee tree (Gymnocladus dioicus), the fruits of which our ancestors used as a coffee replacement. Its foliage is airy so grass grows well beneath it in the summer, and its open, airy branches are interesting in winter.

• Bur oak (Quercus macrocarpa), a relatively fast-growing oak. The bur has large acorns, good for wildlife and craft projects. A spreading tree, bur oak has fiddle-shaped leaves and purple fall color.

- Bald cypress (Taxodium distichum), the tree of swamplands. The bald cypress is also at home in lawns. A deciduous tree that looks like an evergreen, bald cypress foliage is soft and wispy. It can grow to huge proportions and is almost always perfectly pyramidal.

- Cucumber magnolia (Magnolia acuminata) is another huge tree, to 90 feet, with massive leaves. It lacks the shiny evergreen quality of its Southern magnolia cousin and its flowers are more inconspicuous, but cucumber magnolia is hardier in Kentucky and a real looker in its own right.

- American beech (Fagus grandiflora), my favorite tree. The beech is the lord of the forest where it grows; its gunmetal gray bark surrounds a trunk that can grow to massive proportions. Fall color is lemon yellow. Beech is probably the most difficult tree on the list to establish and may be hard to find. You might want to start with a seedling.

- Tulip poplar (Liriodendrun tulipifera), another fast grower. Kentucky's state tree is upright and stately. Its tulip-shaped leaves largely hide its white flowers, but yellow fall color is a good payback. Probably the easiest tree on the list to grow, many school children bring it home after Arbor Day celebrations wrapped in muddy, wet plastic.

- Catalpa (Catalpa speciosa), the cigar tree, named for its conspicuous bean-like seed pods. Lawn neat freaks may not like the seed pods when they fall in the winter, but everyone will love the glorious white blooms in summer and the tropical-looking leaves.

## BACK FROM THE DEAD

The dawn redwood tree, alive at the time of the dinosaurs, was thought to be extinct for thousands of years and then a few specimens were found in China in the 1940s. Back from the brink, the dawn redwood is a good candidate for a place that can hold a big tree—it can reach 90 feet. It is deciduous, with small cones and soft, green needles. Foliage turns reddish in fall, and the bark is handsome in winter. It likes well-drained soil and plenty of moisture.

# Get the blues out of bluegrass

Surely not many are left in the world who believe that Kentucky bluegrass is blue. But there are a surprising number of people, many living right here in this state, who believe that bluegrass is a native, and that vast vistas of grass are the natural condition of the state.

Here's some news: The settlers who first made it through the Cumberland Gap found not fields of green grasses in the central region of what would become the Commonwealth but dense trees, impenetrable cane fields and head-high prairie grasses. In fact, the green lawn grasses we have come to love, including bluegrass, came across that gap in the pockets of the settlers' linsey woolseys. In other words, bluegrass, the namesake of our state, is an import!

What's more, for the last several decades the lawn gurus at the state's universities have shooed us away from planting bluegrass in our lawns. Too high-maintenance, they say. Plant tall fescue instead.

That's too bad, because nothing says go barefoot on the lawn than a stand of bluegrass with its soft, deep green blades. Bluegrass is the first grass to green up in the early spring when anything green is a welcome relief from winter. Bluegrass will also fill in a bare spot better than other grasses because it reproduces from underground tillers as well as seed.

Also, while bluegrass will turn brown and go dormant during a hot, dry spell, it does tend to survive drought better than fescues. "Bluegrass will turn brown, but it will come back," said Newton Sod Farm owner Andy Newton. "Fescue just dies. A lot of people don't know that."

Still, there is plenty of truth to the reputation bluegrass has for being a spoiled lawn child.

Bluegrass does not like the shade very well. Bluegrass succumbs easily to the wear and tear of heavy foot traffic. Bluegrass is prone to turf diseases. And most importantly, bluegrass does not like our typical summertime weather very well. Bluegrass was happy in the England, Ireland, Scotland and northern Europe of our ancestors, where there was plenty of moisture and rarely spells of hot, dry weather. Bluegrass lawns thrive during summers, with abundant rain and cool temperatures. But some Kentucky summers are abusive to bluegrass, and nearly every green and growing thing for that matter. Bluegrass in much of the state can turn brown by mid-June and stay that way until fall.

Fortunately, plant breeders have solved some of the disease problems bluegrass has shown in the past. And new varieties have shown improved ability to withstand heavy traffic. Now bluegrass varieties may be on the way that solve the faint-in-summer-heat problem.

The Scotts Co. of Marysville, Ohio, was the first company to introduce what is likely to be a wave of bluegrass varieties that stand up to summer heat. Scotts' Thermal Blue hybrid bluegrass is part of the company's Heat-Tolerant Bluegrass Seed mixture.

Thermal Blue bluegrass has shown ability to establish rapidly, recover from drought and wear and tear quickly and, most importantly, withstand heat.

Independent tests at Auburn University showed the grass excelled over other cool-season grasses, even tall fescues, at growing in high heat and humidity. The grass also performed well in shade and showed disease resistance.

## Turning the Lawn Blue

The best time to establish bluegrass—or any type of lawn—in Kentucky is Aug. 15 through Sept. 15.

First, get a soil sample done through the University of Kentucky Extension Service. Fertilize and/or lime according to test results.

To seed a small area, loosen the top 2 to 3 inches of soil to provide a resting place for the seed. Broadcast the seed at the rate recommended on the package. Cover the seedbed lightly with straw. To establish seed, it is best to water frequently, rather than deeply. Keep the top inch of soil moist by watering at least once a day until the seedlings have germinated. Once the grass is established, watering can be reduced to once a week if rainfall is sparse.

For large areas of lawn, homeowners may want to rent spray equipment to kill the existing lawn before planting seed. Power seeders that produce good seed-to-soil contact can also be rented. When power seeding into a lawn that has been killed with herbicide, applying straw will not be necessary because the dead grasses will keep the seed in place. Water a large area the same as you would a small area.

While you are watching the lawn seedlings sprout, take a minute to raise the blades on your lawnmower. Grasses compete better with weeds and withstand dry spells better if their stems are left relatively long after mowing. Cut lawns, especially bluegrass lawns, no lower than 3 inches. If the lawn looks dead after you cut it, check the mower blades for sharpness. Dull blades hack the tips of grasses rather than slice them back, leaving a dead brown tip. Sharpen the blades or replace them if necessary.

It is good to know that bluegrass lawns may be on the way back in Kentucky. The Fescue State just doesn't have much of a ring to it.

# Hydrangea, a plant for all seasons

When we were kids, my grandmother had in front of her porch two large shrubs everyone called snowball bushes. We kids hid under the shrubs, jumped off the porch and over the shrubs and dared one another to swat bees off the "snowballs." Those were the days before computers, video games, and 138 TV channels, of course.

When I got a little more horticulturally sophisticated, I learned that the true name for the shrubs we played around was hydrangea. I later discovered, to my horror, that some people call other shrubs—viburnum, for example—snowball bushes. Heretical.

Call them what you will, hydrangeas are excellent shrubs that thrive throughout the state and look great in a foundation planting, shrub border or as a specimen planting. They provide four seasons of garden interest. In the spring, the large green leaves look almost tropical. They bloom in mid- to late summer when most shrubs and perennials are starting to look droopy. Some hydrangeas have excellent fall color, as showy as any scarlet oak or sugar maple. In the winter, some hydrangeas put on a show of exfoliating bark. Flower heads, left on the plant, dry to a crisp brown and can be left to catch the snows of winter or brought into the house for fall and winter arrangements.

The shrub hydrangeas (there is a climbing hydrangea) of interest to gardeners fall into four categories: hydrangea arborescens, or smooth hydrangea; hydrangea macrophylla, or bigleaf hydrangea; hydrangea quercifolia, or oakleaf hydrangea; and hydrangea paniculata, or Peegee hydrangea.

Peegee hydrangea is the least interesting to me. It looks more like a small tree than a shrub, and always seems put in the landscape as an afterthought. Its white blooms come on in June and are arranged in panicles that can be up to a foot long. Nevertheless, Peegee is probably the hardiest of the hydrangea family and will grow in nearly any soil that is not wet.

My snowball bushes were smooth hydrangea. This native species has been planted in gardens for hundreds of years and is an excellent choice for those wanting a "period" garden. Smooth

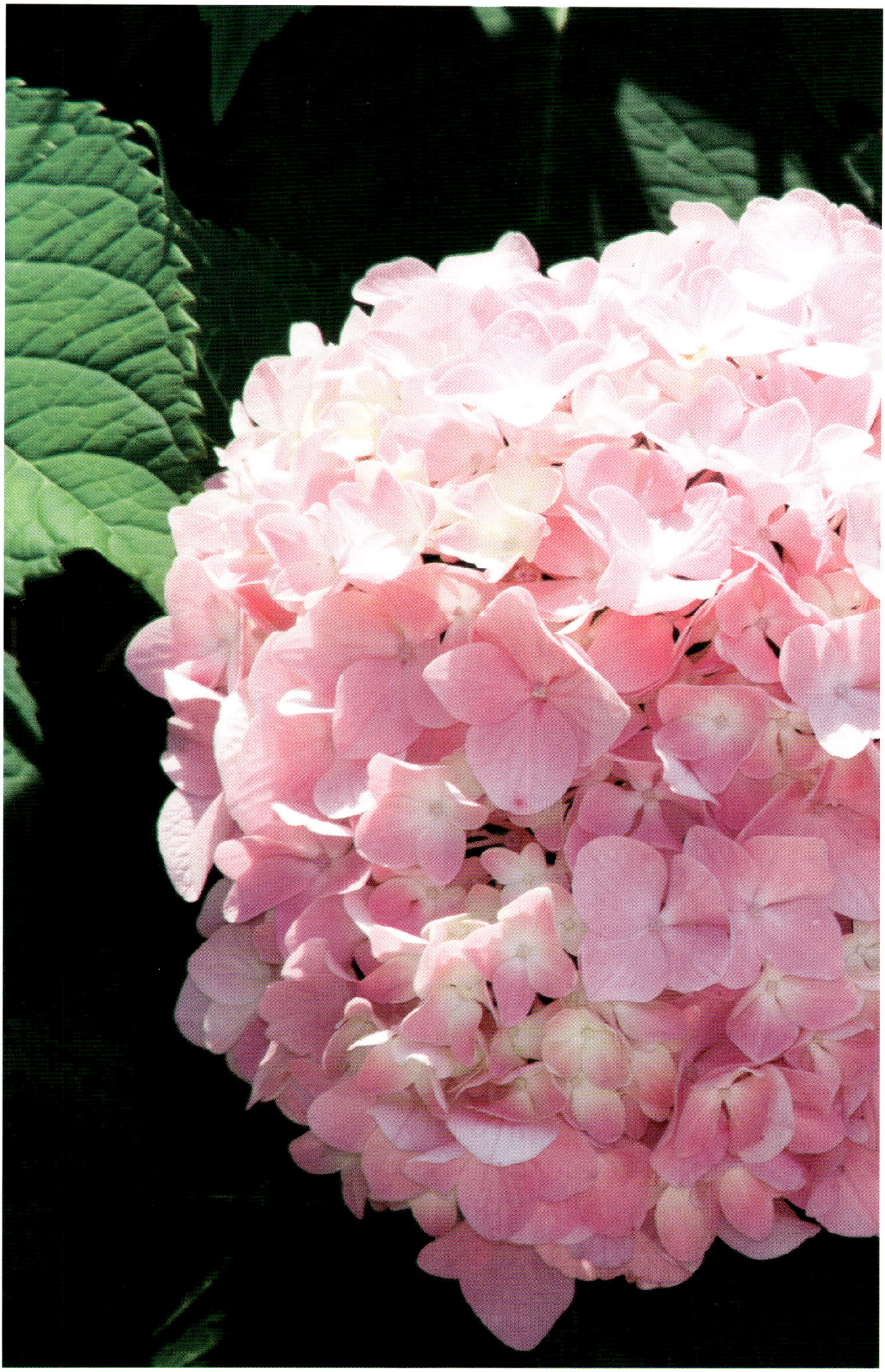

hydrangea's large white mops of blooms start appearing in late June; some are a foot or more across. 'Grandiflora' is the most common variety, but 'Annabelle' is recommended for its more compact size and dark green leaves. Like Peegee, smooth hydrangea is tough, but the massive blooms do have a tendency to flop, especially after a heavy rain; some gardeners plant them in groups of three or more so the plants can support their flowers.

Bigleaf hydrangea is the prettiest hydrangea, with its rich, waxy green leaves and flowers that are pink or blue depending upon the acidity of the soil. Acid soils produce blue flowers; alkaline soils produce pink. Some gardeners will have both colors on the same plant, even the same bloom. To produce deep blue flowers, it may be necessary to sprinkle sulfur or aluminum sulfate around the base of the plant at least twice a year.

The flowers of bigleaf hydrangea come in two forms. The mophead type has large globes of blue or pink flowers. The lacecap blooms are typically smaller than mopheads, with flowers arranged in flat circles or pyramids. To many landscapers, the lacecap types blend in better with surrounding shrubs and perennials. 'Nikko Blue' and 'Forever Pink' are common mophead varieties. 'Blue Wave' and 'Blue Billows' are fine lacecaps.

The bigleaf hydrangeas are probably the puniest of the hydrangeas. Severe winter weather will kill them to the ground. The plant will almost always survive, but it may not bloom after a harsh winter. Also, bigleaf hydrangeas are the least forgiving of areas that drain poorly.

Within the last five years, landscape designers and gardeners have discovered the benefits of the native oakleaf hydrangea. Tree and shrub expert Michael Dirr has written, "The full, rounded-mounded outline, lobed leaves and magnificent white flowers provide full measure for the landscape dollar."

Oakleaf's white flowers not only provide a cool look to a hot, late-summer landscape, but they offer interest as they fade from white to pink or purple. In the fall, oakleaf sports deep burgundy leaves. In winter, the shrub's exfoliating bark is another virtue in the landscape. Like Peegee, oakleaf hydrangea is a tough customer, hard to kill once established. 'Snow Queen,' 'Alison' and 'Snow Flake' are popular varieties.

Hydrangeas are tolerant of a wide variety of soils and growing conditions, but they will do best in a rich soil that drains well. Planting hydrangeas in a wet spot is about the only sure-fire way to kill them.

Though smooth hydrangeas do best in

full sun, the rest of the family likes part-shade conditions. Try to give hydrangeas morning sun and afternoon shade if possible. Spirea bushes make excellent companions for hydrangeas because they will be in bloom about the same time.

Given ample fertilizer and moisture, hydrangeas will fill a space quickly. Check the ultimate size of the variety you are planting and give them plenty of room. Hydrangeas can be pruned, but wait until just after the shrubs have bloomed. Pruning in spring or early summer will eliminate the blooms. If you plant the right hydrangea in the right place, you do not need to prune.

Hydrangeas can be planted in the early spring or in the fall, September through November. Scrounge through garden centers in the fall and you may find some bargains. Then you can have your own yard full of snowballs next summer.

## NEW KIDS ON THE BLOCK

In the decade since the hydrangea article was published breeders have been hard at work creating dozens of new hybrids, many of which bloom all summer. Look for the series Forever and Ever. Warning: In my own garden, they have tended to be punier than some of the older varieties.

# Bernheim Arboretum and Research Forest

When distiller Isaac W. Bernheim gave the state of Kentucky 14,000 acres of forest in Bullitt County, he certainly gave the gift that keeps on giving.

Bernheim Arboretum and Research Forest, more than 80 years old, is more than a woods crisscrossed with hiking trails, though that is a wonder in itself. Less well known is that Bernheim, with its 250-acre arboretum, "green designed" visitor center and more than 8,000 labeled varieties of plants and trees, is a wonderful font of knowledge for gardeners. If you are considering planting a tree, shrub or perennial this fall to replace material damaged in the recent "hurricanes" and ice storms, go to Bernheim. There, you'll likely find a plant to suit your needs growing in the real world, not in the dream world of a catalog.

## A GIFT OF NATURE

Isaac Bernheim, a German immigrant, made his fortune in the distilling industry in Kentucky. When he died, he left the hilly, forested land in Bullitt County to the citizens of Kentucky. The land, remote at the time of Bernheim's giving, now lies within convenient access, just off I-65 at Ky. 245. It is about a half-hour south of Louisville and about a half-hour north of Elizabethtown.

Bernheim's forest has three lakes, 35 miles of hiking and biking trails.

But for gardeners, the real attraction is the arboretum, with its world-famous collections of magnolia, hollies, witch hazel, conifers, crabapples, maples and horsechestnut. Gardeners can see growing side by side species and varieties of plants they may be considering for their own yards. The 8,000 plants in the entire forest and the 3,500 plants, trees, shrubs and perennials in the arboretum represent 120 plant families. Don't go without a notebook!

Besides the collections of woody plants and shrubs, the arboretum at Bernheim has several theme gardens that may inspire gardeners. There are butterfly gardens, cutting gardens, shade gardens and water gardens.

"Bernheim showcases promising plants and displays them so visitors can see them in garden and natural settings," said Dena Rae Garvue, former director of horticulture at Bernheim. "The objective is to identify those plants that do well in north-central Kentucky and to trial plants that are coming into the market."

Garvue said the staff will give a plant a trial of three to five years, and if it proves to be too fussy after that period, out it goes. That's a good system for gardeners everywhere.

Bernheim has also launched an effort to identify plants with exceptional features. Called Bernheim Select, the plants have proven themselves to be winners. The first group evaluated so far is trees that will do well in urban environments. The group of 50 recommended trees can be found on Bernheim's website at bernheim.org. Click on Explore, then Arboretum to find the Bernheim Select list. Flowering shrubs and perennials will be evaluated in the future for outstanding qualities and listed as Bernheim Select, Garvue said.

## Conservation

Bernheim is also about conserving plants and teaching us how to reduce our carbon footprint.

Many types of plant communities, including prairies, swamps and glades, are teetering on the brink of extinction because of development, pressures from agriculture and invasion of exotic species. The horticulture staff at Bernheim is working to conserve some endangered plants as well as entire plant communities. Running buffalo clover, Short's goldenrod, warm-season grasses and the American chestnut are among the plants being grown and studied at Bernheim, Garvue said.

Gardeners can see restored plant communities, including a cypress/tupelo swamp and a prairie. The staff and volunteers are working on restoring a bluegrass savannah and a cedar glade community.

Bernheim uses an integrated pest management approach to controlling weeds and insects in its gardens. Bernheim uses no pesticides except the controls Mother Nature provides, Garvue said, and uses only minimal amounts of herbicides to control weeds.

"As much as we are about plants, perhaps our main mission is teaching people how they can make small modifications in their lifestyle to reduce our carbon footprint," Garvue said. "We're trying to create a space people can be soothed by, excited by and learn from."

As the years pass, the nearly 200,000 visitors who come through Bernheim's front gate can expect to see the forest and the plant communities continue to change, said Executive Director Dr. Mark Wourms.

"It's changed a great deal," Wourms said. "It's a pretty dynamic place. Nature is dynamic, and so is Bernheim."

# Plant a green screen

Ever since Adam and Eve used fig leaves, people have used plants to cover up what they don't want to see.

While we can use fences and walls to block the sights and sounds of an obnoxious neighbor, a rusty tool shed, an outdoor dog pen or all the pipes and electric boxes that sprout on the front lawn, plants are usually a better choice. They are cheaper, less intrusive and less permanent. With a little care and planning, plants can look good in the landscape and do double duty as a privacy screen. Plants, if dense enough or armed with thorns, also can say "keep out" in a way that is friendlier than a "No trespassing" sign.

If you want to hide a view completely or ensure privacy on your side, your best bet is an evergreen screen tall enough and dense enough to do the job. Good choices for Kentucky gardeners would be hollies, taxus (yew), arborvitae, white pine, blue spruce and pyracantha. If you have a problem with deer, you might choose taxus; deer won't eat them.

If you want to soften a view or hide something only during the summer months when the family is outside, deciduous plants may work better than evergreens. The leaves of deciduous shrubs and trees will block views and soften sounds but for only the spring, summer and early fall months. However, if planted densely enough or in layers, deciduous shrubs can be almost as effective as evergreens in providing a screen. A row of tall viburnum shrubs fronted by an offset row of shorter viburnum species, for example, will block views for at least nine months of the year and soften the view for the remaining months.

Another choice for most but not all of the year is ornamental grass. Ornamental grasses, such as switchgrass or miscanthus, grow quickly into view-blocking backgrounds. When the frost hits, the grasses turn shades of red, yellow and brown. But the grasses should be cut to the ground in March. So between March and May, they will not perform their screening duties.

If you have a yard neighborhood kids use as a cut-through or that is the favored potty spot for pets, you may want to

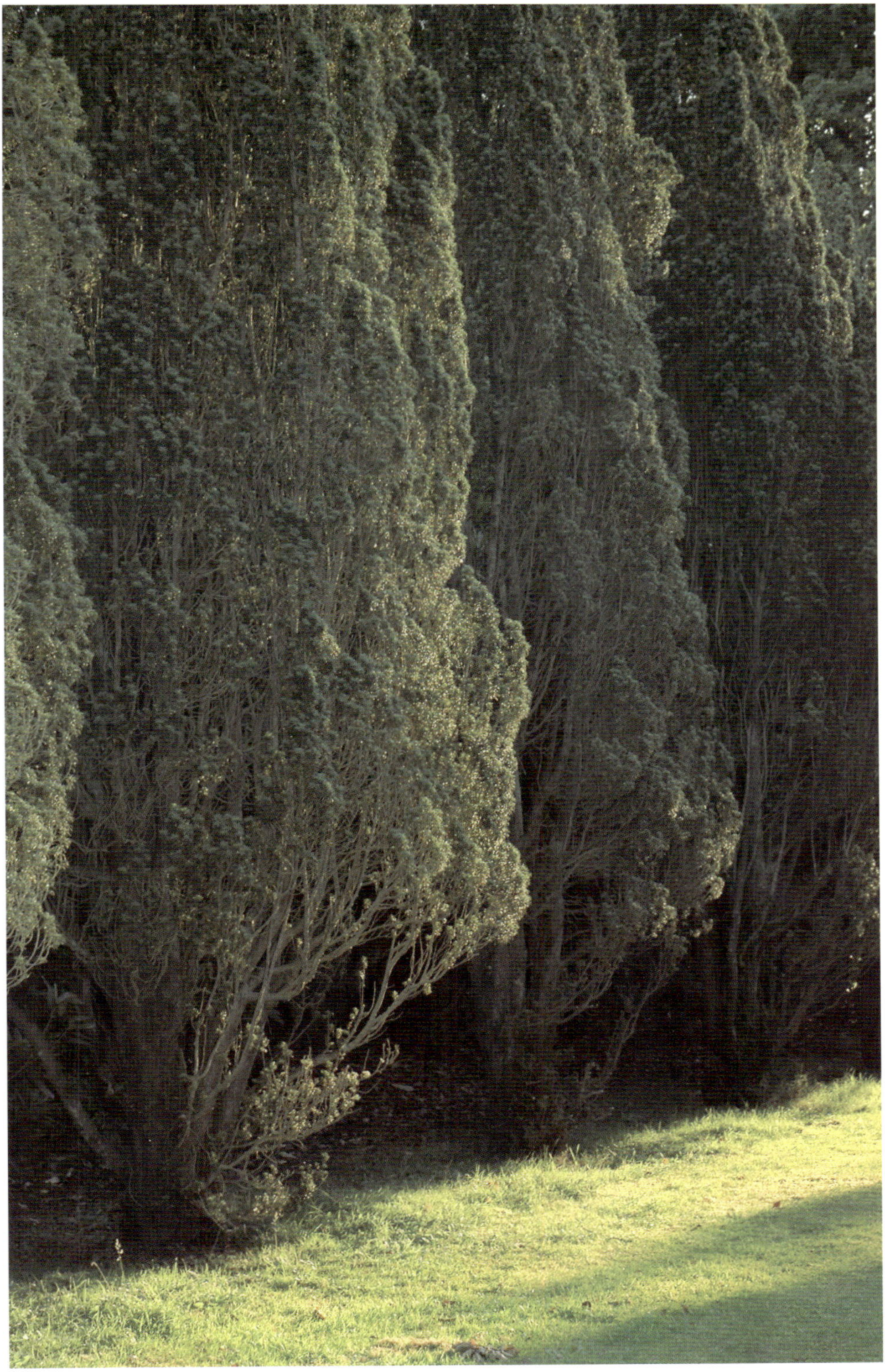

create a screen that no kid or pooch wants to push through. The answer is plants with spines or thorns. Firethorn (pyracantha) is a good choice. Planted closely together, firethorn shrubs form a dense, thorny barrier quickly. The plant, with its red, yellow or orange berries, is also attractive.

## Fitting in

A good screen or barrier fits into the landscape and does not appear to be there just to hide something. Developers frequently plant four or five exotic evergreens around a cable box that stick out like sore thumbs in the landscape.

One way to blend your screen with the landscape is to add elements of the landscape to the screen or use the screen as a backdrop for a perennial or annual flowerbed. Roses, daylilies, peonies and other perennials planted in front of taxus or arborvitae add color and interest, while the green background sets off the beauty of the blooms.

Another technique is to create a berm planting that serves as a screen/plant bed combination. The taller plants can be toward the back with the shorter perennials, annuals and bulbs in the front.

## Additional advice

Be careful where you dig and plant. Cable boxes and pump lift stations have underground wires. Know where they are before you start digging. Also, the boxes and pipes may need access occasionally. Don't put in plants that are impenetrable to a maintenance person.

Be nice. You may not like your neighbor and may want to screen him from view, but you don't want a lawsuit, either. Don't plant trees that may fall on his property or invasive plants, such as bamboo, that may overrun his yard.

Don't scrimp on size and quantity. A screen should hide the view completely and quickly. Buy larger rather than smaller specimens and plant them more densely than you ordinarily would. You can always cut something out later if the screen grows too dense.

Plant mixed screens, if possible. A row of the same species invites trouble; if disease gets a foothold, the whole screen goes. Mix different types of evergreens or evergreens and deciduous species.

I am not crazy about shearing or clipping shrubs, but if you want a screen or hedge cut to shape, keep the plant narrower at the top than at the bottom. That way, sunlight will reach the lower branches, keeping them full and leafy.

## LEYLAND CYPRESS

Many fast-growing plants are undesirable, but one of the good guys is the Leyland cypress. An upright evergreen, it can grow up to 60 to 70 feet in a few years. 'Castlewellan' is a popular selection that has golden new growth; 'Emerald Isle' has bright green foliage and is shorter, up to about 25 feet. Be careful, though, it requires occasional pruning to keep it in check.

# Foreign invaders attack our state's forests

Here we go again.

Yet another Asian invader is threatening to decimate the state's forests, and when its work is done, we'll likely be all the poorer for it.

The emerald ash borer was discovered first in Shelby County in the spring or 2009 at a camp near the Franklin County line. A few days later another infestation was found on a homeowner's lot in Jessamine County. What that means is that probably, within the next decade, the state will lose most of its ash trees, a tree that makes up anywhere between 10 percent and 17 percent of the state's forests.

At this writing the bug has been found in counties in north-central Kentucky, including Franklin, Fayette and Jefferson, and 29 counties are under a quarantine. That means firewood can't be taken into or out of those counties, ash trees from those counties can't be sold as nursery stock, and ash lumber must be inspected and debarked before it is taken out of the 29-county region.

Valiant as the effort is, even state entomologist John Obrycki admits it will likely do little good to contain the spread of the borer. The quarantine is almost impossible to enforce, and the deadly little borer can fly.

The ash borer was first introduced into this country likely from a load of pallets shipped to Michigan. In just a few years, the borer has destroyed tens of millions of ash trees in the upper Midwest and is in every state surrounding Kentucky. The borer kills green, white and blue ash trees by feeding on the inner bark of the tree, cutting off water and nutrients to upper branches. The feeding usually starts in the top of the tree and works its way down. So by the time the homeowner may see the distinctive D-shaped exit holes, the borer has done enough damage to ensure the tree's death within three to five years.

That's sad because Kentucky's forests contain some 200 million white and green ash trees. The ash is a relatively quick-growing tree, well-shaped and sporting beautiful fall color. Its wood is sometimes used for furniture, but it is best known as the ingredient in wooden baseball bats. Nurserymen had developed several ash varieties over the years that simply glowed orange and purple in the fall. But many nurseries in the state that stocked the tree have plowed them under because no one will buy them, given the threat of the borer. The city of Lexington, for example, stopped planting ash as street trees because of the threat of borer damage.

The emerald ash borer joins the wooly adelgid, another invader from Asia, as threats to our state's forests. The wooly adelgid attacks hemlock trees by feeding on the needles and injecting a toxin into the tree. Hemlock trees thrive primarily in the mountains, where they not only add beauty, but their ferny evergreen foliage creates the shade that cools mountain streams, and their root systems filter pollutants out of groundwater while holding the soil in place.

There is not much the state's gardeners can do to protect against these insects on a scale the size of the state's forests, but there are some remedies to protect a valuable tree in the home landscape.

The wooly adelgid can be sprayed in late March through April with insecticidal soaps that effectively smother the eggs in the furry sacks that give the insect its common name.

Both hemlocks and ash trees can be treated with a systemic insecticide to fight off the insect. The chemical imidacloprid spread around the drip line of the tree and watered in will poison the insect as it feeds upon the tree. Imidacloprid is sold under several brand names, including Bayer Advanced, Merit and Marathon. A once-a-year treatment in spring is likely enough to protect the tree. If you use a systemic insecticide to protect your ash or hemlock, do not use it around open water because it can kill aquatic plants and insects. As with any chemical, use only according to label directions.

Meanwhile, gardeners also should avoid buying ash trees for the home landscape. Good replacements would be the red or sugar maple, swamp white oak, willow oak, or hickory.

Someday, we will breed trees resistant to these Asian invaders, and natural predators will evolve to feed upon them, reducing the threat to our forests. In the meantime, we will have to sit and watch another ecological disaster hit our forests, maybe one not as bad as the chestnut blight or Dutch elm disease, but certainly one our forests do not need.

## SUCCESS STORIES

We may finally be winning the war against the pestilences that decimated our eastern forests—chestnut blight—and our streetscapes—Dutch elm disease. Scientists have crossed American chestnuts and elms with related species that are disease resistant to produce specimens that have 90 percent or more of the original genes but with built-in disease resistance. These trees are now being planted all over Kentucky and are being watched carefully for signs of infestation. Maybe our grandchildren will get to walk among forests of giant chestnut trees and drive down streets lined once again with elms.

# Winter

# New Yorkers high on Kentucky gardener Jon Carloftis

Let me tell you about a Kentucky gardener and landscape designer who has literally scaled the heights of New York City, designed gardens for the Big Apple's glitterati and found his work splashed all over the country's gardening and home decorating media.

Born in Rockcastle County, one of six children, Jon Carloftis grew up without a telephone until the late 1960s; the family did not get a television until 1985, when Carloftis had already left for college.

Carloftis got his first horticultural education walking through the woods with his father to fix the family's water line. The elder Carloftis would name the trees of the woods for his son.

"As a kid, that didn't mean anything, but it must have sunk in," Carloftis said.

Carloftis later studied horticulture and art history at the University of Kentucky while earning a degree in communications. In 1988, Carloftis moved to New York, where he befriended the doormen of posh apartments and condominiums on the upper East Side. They agreed to hand out the cards he had printed up: "Jon Carloftis, Rooftop Garden Designer."

"I had never been on a rooftop before in my life," Carloftis admitted.

His first job was for a woman who wanted her extensive modern art collection mirrored by the garden Carloftis was to create outside on the rooftop.

"I want you to feel this art and then take it outdoors," the woman told him.

Carloftis created a garden full of plants that matched the "edginess and aggressive sophistication" of his client, and soon word of Carloftis' work spread. He found himself designing rooftop gardens for dozens of other well-heeled clients. Soon he had about 30 clients in New York and other places around the country. Carloftis moved to Bucks County, Pa., where he and helpers gather plants and commute to the city to work with his clients.

Carloftis said he works only with clients he "can get along with."

"I have to form a relationship with the client," Carloftis said. "I have to like them before I can make them a beautiful garden."

Since he made that first rooftop garden in the late 1980s, Carloftis' work has been featured in several home and gardening magazines, including *Metropolitan Home*, *Garden Design*, *Country Home*, *Country Garden*, *House Beautiful*, *Martha Stewart Living* and *Country Living*. He has made appearances on HGTV, Martha Stewart's show and Style TV.

## Unpretentious Gardener

Rubbing elbows with the rich and famous has not erased the Kentucky mountains from Carloftis' lecturing, writing or gardening style.

In Carloftis' books, *Secrets of a Kentucky Garden* and *First a Garden*, Carloftis urges readers to create their own style, one that feels good to them. On a trip back home, he told his Kentucky audience that gardeners should plant what they enjoy and use what decorations they find attractive.

"If it costs $5 and you think it's beautiful, then use it," Carloftis said. "It's your house and your life. Do what you like."

Carloftis said movie stars and other famous people are like everyone else in that their gardens all have a view they want to hide and a view they want to accent. He recommends espaliered apple trees to hide an ugly wall; junipers to hide "an ugly driveway or ugly neighbor."

"We all have ugly things in our life we want to hide," Carloftis said.

## CARLOFTIS' 8 WINNERS

While Carloftis' books are short on must-dos and must-haves, he does recommend annuals, perennials, shrubs and trees in his book, *First a Garden*. These plants have worked well on the rooftops of New York, where they are exposed to the extremes of the elements.

"If they do well in New York, they'll do well in Kentucky," Carloftis said.

Here are his top eight picks in each category:

**White annuals:** Cosmos, sutera (bacopa), licorice vine, mission bells, cleome or spider flower, nicotiana or flowering tobacco, dusty Miller and impatiens.

**Colorful annuals:** Nasturtium, verbena, dahlia, salvia, vinca, plumbago, French marigold and coleus.

**Perennials:** Sedum, maidenhair grass, coneflower, summer phlox, daylily, hosta, iris and black-eyed Susans.

**Trees:** Flowering dogwood (cornus florida), hemlock, willow oak, Eastern redbud, magnolia, Japanese cedar, Japanese maple and umbrella pine.

**Shrubs:** Robusta green juniper, Japanese andromeda, American boxwood, hydrangea tardiva, dwarf hinoki cypress, boulevard cypress, chaste tree and burning bush.

# Bark up the right tree

When gardeners shop for trees to decorate the home landscape, they look for size and shape, shade potential, fall foliage color. Understandable. But there is another dimension to the beauty of trees we ought to consider lest we be guilty of not seeing the forest for the leaves.

I am talking about bark.

The bark is the tree's skin. It serves the tree as protection from the elements just as the roof over your head shelters you. But bark can also be a thing of beauty in and of itself.

Stand back and look at a group of trees in the woods or on a hillside in summer. All you see is leaves.

Look at the same scene in winter from a distance, and all you see is drab gray. But walk closer. Notice how the Kentucky coffee tree has scaly, recurving ridges.

Touch the bark of the American beech to commune with its smooth grayness. Stick your fingers in the smooth, patterned ridges of the bark on a black walnut or red oak. In short, experience the understated beauty of tree bark.

There are trees that need no close examination to appreciate the beauty of their skin. They have bark that is more than decorative in its own right; in fact, the bark may be the most decorative aspect of the tree. Anyone who has seen a hillside of white birch in New Hampshire on a sunny winter's day knows what I mean. Or, closer to home, if you come upon a hillside of predominantly beech trees in the fall, yellow leaves aglow setting off the gunmetal-gray bark, you have found one of nature's most fabulous, if ephemeral, scenes.

Another example is the sycamore in winter. I would never recommend you plant a sycamore tree—its branches are weak, its roots seek out water and sewer lines, and its large leaves are annoying to rake—but the sycamore in winter, lining a creek as it usually does, is a standout. The sycamore's snowy white top branches contrasted against a gray winter sky are a delight in a sometimes dismal season. Ask your neighbors who may not be aware of the sycamore's bad habits to plant a few sycamores for your viewing enjoyment.

## Beautiful bark

Trees with striking bark come in many sizes, shapes and forms. Some trees, like the beech or birch, have bark of a strikingly different color and texture. Some trees have mottled bark of two or more colors. The lacebark elm, for example, has splotchy bark of a green and pink shade; a fabulous example grows on Main Street in Frankfort, just around the corner from Liberty Hall. Some trees—the paperbark maple or river birch, for example—have peeling bark. The peeling bark of the paperbark maple reveals a cinnamon brown undercolor that is a standout in a winter snow scene. Some trees, like the shagbark hickory, have splitting bark that adds a jagged edge to the form of the tree.

Below is a list of trees, not mentioned above, with striking bark along with a description of their growth habits. All can be grown in Kentucky but some, as noted, are more of a challenge than others:

- White oak (quercus alba)—a huge, slow-growing tree with, as the name suggests, a very light-colored bark. It's the tree suitable for making whiskey barrels but not a good candidate for the average yard because of its size.

- Crape myrtles (Lagerstroemia indica)—are more often large shrubs than trees in Kentucky and are grown primarily for their summer flowers, but crape myrtles have wonderfully smooth red, brown and yellow patterned bark.

- Sourwood (Oxydendrom arboreum)—has blocky, patterned bark. The tree is interesting all year round with its summer blooms and scarlet fall foliage. But it demands well-drained, acid soil and is probably best grown in the southeastern part of Kentucky.

- Cucumber magnolia (magnolia acuminata)—tree lacks the flashy flowers of its family members but is a tough, easy-to-grow tree with bark that sports a ridged pattern.

- Franklin tree (Franklinia alatamaha)—Ben Franklin's namesake—is a small, ornamental tree with summer flowers and smooth, gray bark broken by irregular vertical fissures. Can be tough to establish but worth the try.

- London plane tree (Platanus xacerfolia)—a sycamore hybrid that is better behaved than its parent. It has peeling bark that reveals a green, rather than white, undercolor. Good street tree.

- Japanese stewartia (Stewartia pseudocamellia)—a mid-size tree with small, white flowers in summer and bark similar in color to the lacebark elm but with a peeling effect. Appreciates afternoon shade.

- Lacebark pine (Pinus bungeana)—a large evergreen tree with exfoliating bark that is sometimes whitish gray, sometimes greenish-brown. Grows slowly and can be hard to find at local nurseries.

- Katsura tree (Cercidiphyllum japonicum)—a tough, mid-size tree with large, greenish-blue leaves. The bark of the katsura tree is slightly shaggy and rich brown.

- Amur maple (Acer ginnala)—has nearly the same smooth, gray bark as the beech but is a small tree with brilliant red fall color.

This fall, when you are going to the nursery to buy your tree (you are planting a tree every fall, aren't you?), take this list with you and consider buying a tree for its bark as well as its foliage and flower.

Remember that leaves and blossoms are fleeting, but the bark is always there.

## WATCH THAT TRIMMER

String trimmers are great tools, but they can be death to trees. Always keep that whirling line away from the bark of trees. Injury to bark allows insects and disease to gain a foothold. Probably more trees are lost to string trimmer injury than to storms. Use mulch—not too deep—around trees to avoid having to get too close to trim.

# Is your garden for the birds?

In the nature center at Bernheim Forest, lying on a bed of cotton under a glass case, is one of the last of the passenger pigeons. The bird is a reminder that a species that once numbered in the billions can cease to exist and a warning that many bird species are on the brink of extinction.

Some of our most colorful birds, some of our best songsters, some of our best insect eaters are declining in population. Studies done by Cornell University and other schools show that the numbers of many species of warblers, flycatchers, tanagers, thrushes and other songbirds have dropped by nearly half within the last three decades.

While much of the decline in bird numbers has to do with the loss of tropical rainforests, habitat loss here in our country also gets much of the blame. We gardeners cannot do much single-handedly to save the jungle, but we can do a great deal to create a friendlier habitat for birds in our backyards.

While planning a garden, making improvements or just adding another plant or two, think about putting in plants that help the birds. Many of the plants that are perfectly acceptable additions to the landscape also make good habitat for birds. We just have to give them a little thought when making a trip to the nursery or garden center. Some garden centers even help out by indicating on the tags which plants are bird magnets.

Remember that in planting with birds in mind, the goal is not so much to attract numbers of birds as to offer habitat suitable to the widest range of species. After all, if it's just numbers we want, we could pour out a bag of corn on the ground and watch the pigeons, sparrows and starlings come flocking by the thousands. What we want is feathered diversity.

The best way to achieve that is to create plantings that offer at least two of the things birds seek out: water, food, shelter and nesting sites.

We can feed the birds artificially, of course, with feeders. But plants can offer food, shelter and nesting sites all at the same time. Plantings that are attractive to the widest range of bird species will use a combination of trees, shrubs, perennials, annuals and vines.

Some trees that attract birds and at the same time are attractive to us include: birch, cherry, plum, crabapple, flowering dogwood, Chinese dogwood, red-osier dogwood, eastern hemlock, green ash, hackberry, mountain ash, Japanese maple, serviceberry, sweetgum, red and pin oaks, and tulip poplar.

Shrubs may be the best bet for attracting birds because they can offer food, quick escape from predators, and a place to nest. Most bird species nest within 5 to 8 feet of the ground. Some shrubs that will bring in the birds are: sumac, blackberry, elderberry, Yaupon holly, juniper, spicebush, viburnum and witchhazel.

Many birds eat the seeds of perennials, nest among them or use their dried stalks for nesting materials. Some perennials recommended for birds are: aster, black-eyed Susan, brown-eyed Susan, coreopsis, goldenrod, globe thistle, pinks, scabiosa, purple coneflower, statice and grasses, including little bluestem and bulbous oatgrass.

Most of the annuals we plant have

been bred to offer us a blast of color in the summer and do not have much to offer birds. But a few are attractive to our feathered friends for the seeds they provide: amaranthus, bachelor button, California poppy, gloriosa daisy, marigold, sunflower and zinnia.

Finally, several bird species are fond of nesting among vines. If the vines offer fruit, so much the better. Some vines that attract birds are: bittersweet, English ivy, fiveleaf akebia, grape, pyracantha, trumpet honeysuckle, and Virginia creeper.

Gardeners who have the space to put together one or more species from each of these groups have gone a long way to offering birds a snack as well as a safe haven.

A good example may be a garden with a serviceberry tree and two viburnum bushes, with New England asters and coreopsis at the base of the viburnums. The serviceberry will provide fruit in the summer, the viburnums will provide fruit and a place to nest in the fall, while the asters and coreopsis will offer seed in the fall and winter.

If you have a small garden next to a fence, you may want to plant pyracantha near the fence, purple and white coneflowers in the front of the vine and scabiosa in front of the coneflowers. The pyracantha will offer shelter and orange berries in the fall; the coneflowers and scabiosa have seeds that finches will flock to in late summer and fall.

Finally, if you can provide water in the form of a bird bath or flowing water along with the right plants, birds will find your place irresistible. Many uncommon species of birds will avoid coming near human dwellings, but they are driven to the sound of flowing water and will forget their shyness to drink and bathe. If you have access to electricity outside, fountains and pumps can be put together inexpensively to provide flowing water that requires little or no maintenance.

To learn more about gardening for the birds, you might want to check out University of Kentucky wildflower expert Dr. Tom Barnes' book, *Gardening for the Birds*. It is available from The University Press of Kentucky.

## BACK OFF, BIRDS

Most birds are good in the garden, most of the time. But some species are a nuisance, and when the fruit is ripening, even benign species like robins and mockingbirds can be unwelcome. To keep birds out of fruit, such as cherries or blueberries, buy netting that can be placed over the plants. Garden centers also sell shiny tapes, scary eyes and other products to deter birds. A homemade bird frightener is a shiny pie pan on a string and tied to a stake. Birds don't like the flashes of light and the movement of the pans.

# Catch up on your reading

As a garden writer, I get dozens of gardening books every year from publishers hoping I'll pass the word along to my readers. I read them all but mention only a few every year because so much of the gardening literature put out has little more to offer than what has come before.

I look for three qualities in a garden book before I pass along a title to others: information, perspiration and inspiration. I want to find information in a book that will be useful to the veteran gardener as well as the newcomer. I want it to be easy to understand and in a logical format. I want it to be horticulturally accurate without being overly scientific.

I want a garden book that tells me how to build a raised bed, properly plant an evergreen, cure onions or landscape a new property without telling me there will be no sweat or manual labor involved. If an author tells me I can have a garden that is the envy of my neighborhood with only half an hour a week investment, I set his book aside.

Perspiration—just tell me how much.

Finally, I want a book that inspires me. Gardening is work (see above) unless you are well off enough to hire your help. So write me a book that moves me to want to work out in the summer sun in July and August when every other self-respecting soul is holed up in air conditioning. The writing should be crisp and clear; the pictures should be abundant. Here are four books that meet the test and provide more than enough reading to keep you busy until time to pick up the hoe.

*Flora: A Gardener's Encyclopedia* is a massive tome that comes in two volumes—two very large volumes. With this book you can read and exercise at the same time. *Flora* does not list every plant in the world or every cultivar, but it comes close. More than 20,000 plants are described in the book.

Before the A through Z encyclopedia of plants begins, *Flora* has an introductory chapter on the plant handiness zones. Another chapter covers the plant groups—trees, shrubs, grasses, etc. The first is interesting reading, but for the seasoned gardener, it's nothing really new. The plant group chapter seems superfluous.

The encyclopedia does list plants in alphabetical order by Latin name, and that may throw some beginning gardeners. But the listing gives, in cryptic form, complete cultural information. The pictures are absolutely stunning. In the encyclopedia, mixed in among the usual small pictures are larger pictures of plants in their natural settings. That makes it much easier to envision what a plant will look like in a garden setting.

Another encyclopedia-type garden book, *The Southern Living Garden Book*, focuses on plants of the South.

The book's strong points are its plant selection guide lists in the front of the book, lists for everything from landscape plants with fragrant flowers to plants that do well in the shade. As you might expect from a publication put out by *Southern Living* magazine, the more than 1,300 photos are uniformly excellent.

On top of that, Steve Bender, who edited *The Southern Living Garden Book*, is one of the best garden writers in the country. He not only knows what he's talking about; he clearly is having a good time writing about gardening.

*Restoring American Gardens* is one of the better books around on the hot gardening trend of planting heirloom plants. Many gardeners have old houses they have restored authentically inside and out; they want their gardens to match the home's time period.

*Restoring American Gardens* divides the country into regions and talks about garden styles to match the history of those places. The author devotes a great deal of space to gardening and architecture, a topic most garden writers eschew because they may know a lot about gardening but precious little about architecture.

The last part of the book contains an encyclopedia (that word again!) of heirloom plants and, most helpfully, the book has lists of sources. There is nothing more frustrating than deciding you want to try old-fashioned plants and not being told where to find them.

Finally, give *Shaker Medicinal Herbs* a read. The book is a fascinating journey into history, culture, horticulture and medicine. You learn a little bit about the world of the Shakers of Kentucky, Ohio and New England, but more about how they grew not only their own food but also their own medicines. Shaker herbal cures may make us cringe, but they were medical state of the art for much of the 19th century.

## THE RIGHT ROSE

I recently read *The Right Rose in the Right Place* by Peter Schneider (Storey Publishing www.storey.com), an interesting book for rose lovers. Schneider's thesis is that with careful choosing of varieties, anyone can grow roses anywhere. He has an extensive list of rose favorites, and the photos are gorgeous. Some of his recommendations may be hard to find, however.

# Match the garden to your cottage

A few years back, a woman called me, having seen my column in *Kentucky Monthly*, and wondered if my garden could be included in her group's garden tour.

I think I mumbled something about being fresh out of tea and cucumber sandwiches, or maybe not that, but somehow managed to persuade her the group really didn't want to stop and see my garden.

If I had told the truth, I would have said my garden's expression of my skills is of the "do what I say, not what I do" variety. In other words (sorry to tell you this, trusting readers) my garden is frequently a source of scandal to what I know a good garden should look like. Colors clash. Tall plants grow where short ones ought to be and vice versa. Variety labels get lost. Roses get black spot. Weeds sneak in. And something, somewhere, is always dying. Not usually a pretty picture. The ladies wouldn't have been amused.

For a long time, I thought my garden was just a reflection of my frequently disorganized self. I meticulously plan my work—writing and teaching—but when it comes to my hobby, gardening, I like a more laid-back approach. Another explanation—excuse, I suppose—is that because I am a gardening writer, I have lots of friends who give me garden plants. And who is going to refuse free plants? Plus companies pass along plants for me to try out in my garden. Every year, for example, a couple of rose companies send me their latest hybrids to test—and frequently to kill. So, it's hard to have a color-coordinated, planned-to-the-letter garden when I'm always sticking in somebody's gift.

Happily, a few years back, I discovered a name for the type of gardening I do. It's called "cottage gardening."

Cottage gardening is a rather carefree style of gardening—there are a few rules we'll get to in a moment—that has been practiced in rural England for more than 300 years, and it is still common there, although much less common here in the States. Cottage gardens usually appear at the front of rural dwellings, though they are frequently found in cities as well, where they take up space between the curb of the street and the front porch.

Cottage gardens include a riot of flowers—annuals, perennials and biennials, with little or no regard to color; shrubs; some vegetables and herbs; and even some fruits, including strawberries and raspberries. They require little maintenance because the flowers are typically packed in rather tightly and choke out weeds. Something is always blooming in the cottage garden, and if your garden friends pass along some seeds or cuttings, there is always a place for them.

Warning: If your idea of the picture-perfect garden is a row of taxus lined up under a windowsill like soldiers on parade, then the cottage garden is not for you and you're allowed to stop reading here. But if you like things a bit more informal, like a garden that always has something in bloom—and occasionally something to eat—and enjoy a garden that needs little to no spraying for insects and disease, you might want to try designing a cottage garden this summer and start planting this fall.

## Try one?

While the great thing about cottage gardening is that there are few rules when it comes to design, the gardener cannot escape the essentials—the right plants have to go in the right place. In other words, if the cottage garden will get full sun, then shade-loving annuals and perennials are going to pass out there; and if the garden is in the shade, the sun worshippers will grow but refuse to bloom. Soil type and drainage also are critical. Often, gardens planted near the foundation of a home drain poorly, and only plants that tolerate wet feet will perform well. The drainage problem can be solved by raising the height of the planting area with additional soil and compost.

Plant freely and abundantly. The ideal cottage garden keeps the taller stuff—like foxgloves, shrubs and vines—in the back

## WHAT PLANTS WOULD FIT THE BILL?

Here are a few suggestions:

**Annuals:** red spider zinnia, nasturtium, painted tongue, 'Oxford Blue' love-in-a-mist, flowering tobacco, strawflower, 'Blue Boy' bachelor button, love-lies-bleeding, cockscomb. Many of these annuals will self-sow and come back every year.

**Perennials:** sweet peas, 'Festiva Maxima' peony, Siberian iris, German bearded iris, delphinium, bee balm

**Biennials:** 'Giant spotted' foxglove, 'Nigra' hollyhock

**Bulbs:** daffodils, grape hyacinth, species tulips

**Vines:** Moonflower, hyacinth bean, cardinal climber

**Old roses:** Indigo (1830), Portland (1837), Charles De Mills (1746), Bella Donna (1844)

**Shrubs:** Lilac, daphne, forsythia, hydrangea, viburnum

**Heirloom vegetables:** 'Listada De Gandia' eggplant (1850), 'Amish Deertongue' lettuce, 'Brandywine' tomato, 'Large Red' cherry tomato, 'Red Fig' tomato

**Herbs:** hyssop, English lavender, thyme

and a climbing rose if there is room. And remember that the cottagers frequently venture into the garden to gather herbs for the kitchen, so include a few of those, along with some vegetables that fit the scale of the garden, perhaps tomatoes and peppers.

If I were starting a cottage garden from scratch, I'd select heritage plants that grew in the typical cottage gardens in England and the U.S. during the Victorian era. That was a time when not only cottage gardening but also interest in gardening and plants in general reached its zenith.

and the shorter plants toward the front, but that's about the only rule on the order of placement. Keep it a riot of color and texture. And let the annuals that self-sow spread their babies about and come up where they will. A dense planting will reduce the need for weeding.

Then you have to find the plants. You will want a mixture of shrubs, annuals, perennials, maybe a vine or two winding up a trellis (moon vine is a good choice)

## WATCH THOSE WINDOWS!

When choosing plants for your foundation, pay careful attention to the ultimate size of the species. You don't want plants that grow so tall and bushy they cover up the windows of your house. This would be bad for a couple of reasons: one, you want to see outside; two, plants give potential burglars a good place to hide while they jimmy the windows.

# The school of hard knocks

When I was still 20-something, I lived in Dayton, Ohio, where I wrote a garden column for a local magazine. An area garden club invited me to speak about organic gardening or something or another, which I agreed to do. When the president of the club called prior to the meeting to ask for my bio to pass around to the members, she was horrified to learn that I didn't have a degree in horticulture. Her tone of voice told me she just knew the ladies (and one man, it turned out) weren't going to learn anything at that meeting.

I don't remember much about what I said at my talk or whether the reception from the members was warm or chilly, but I do know I felt then, and still do, that, in the garden at least, book learning doesn't go very far. If you want to really learn about gardening, you have to plant your feet in the soil and get your hands dirty.

Having said that, I am going to pass along a few lessons I have learned from getting dirty in the garden. And when you're finished reading them, put down the book and get outside in the garden.

## If at first you don't succeed

Too many would-be gardeners give up on gardening because their plot doesn't come out picture-perfect on their first effort. Whether you have a garden of flowers or vegetables, it is going to take a while to learn the vagaries of your space—soil, sunlight, drainage, frost-free dates, etc. In fact, plan to spend several years learning. And before you know it, your garden will be productive—if you keep trying. And here's another dirty little secret: Your garden will never be finished to your satisfaction, and there will always be something to change, take out or improve. That's the fun of gardening.

## Let the strong survive

This advice may appear to contradict the above, but it doesn't really. Some plants just aren't meant for your garden space, either because they don't like the soil, don't like the drainage or, I'm increasingly led to believe, they just don't like you. And there's no reason to devote hours of spraying, pinching, manuring

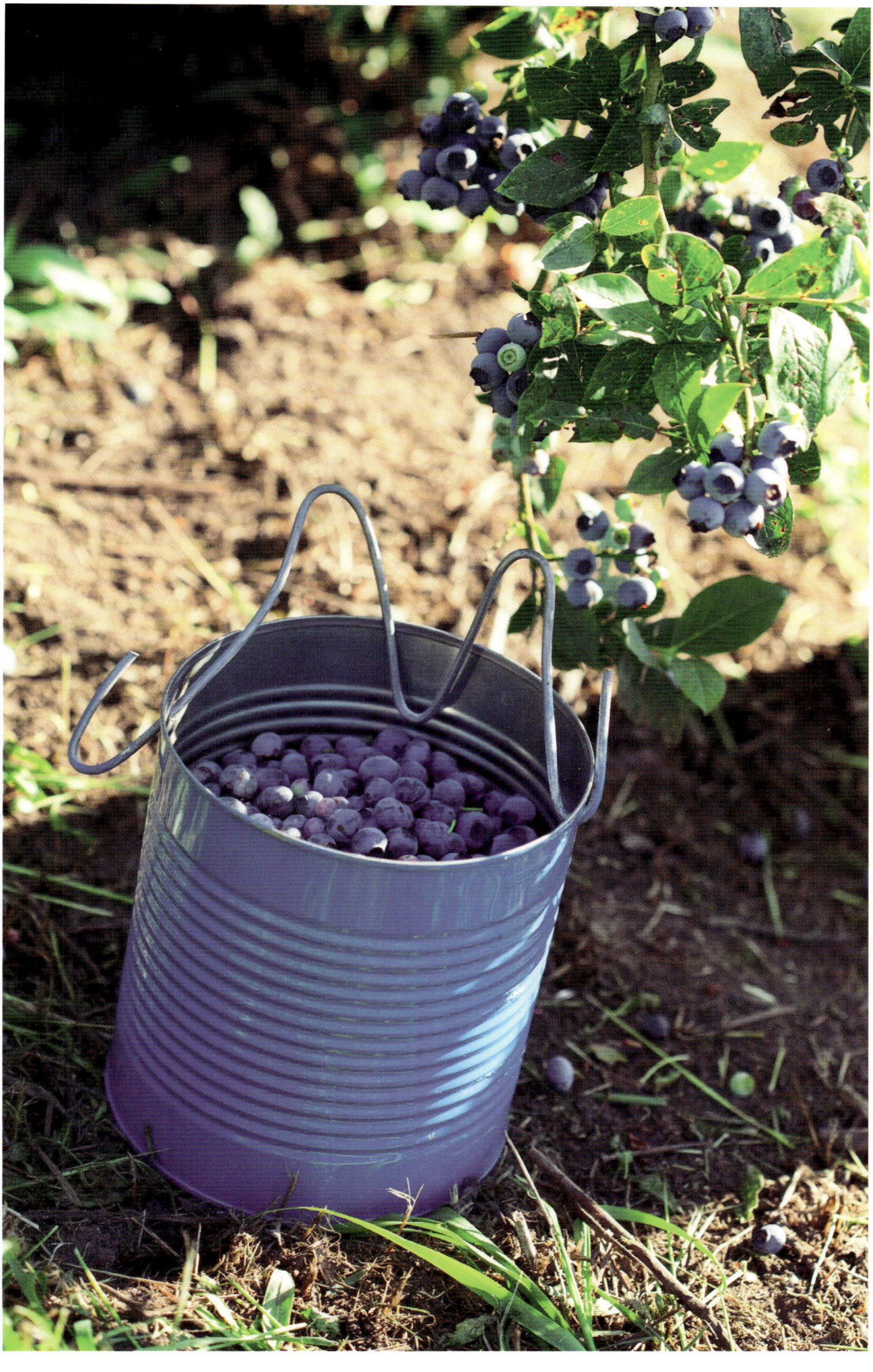

and pampering something that just won't grow for you.

Take my relationship with the rhododendron family, which also includes what we call azaleas. Rhododendrons are magnificent; azaleas are stunning. But they hate me. They won't grow for me. I have killed dozens over the years trying to get just one to grow, settle in and bloom like the ones in catalog pictures. I have tried every trick in the book. Nothing. Ditto blueberries. I've killed off a fortune in blueberry shrubs before they even yielded up one little, hard, bitter fruit. I was determined to grow them again because they are supposed to ward off senility or Alzheimer's disease or I forget what else. In other words, there comes a time to give up on certain plants.

## A Stitch in Time Saves Nine

A great expression from the days when people actually knew how to sew, it means action now prevents more work later. Nothing is as true in the garden as that lesson. Is your garden a patch of weeds bigger than you are right now? If so, you probably didn't rogue out the little boogers when they were small enough to pluck with your fingers. Or you didn't put down mulch as I have told you. Did your garden drown in the spring rains, causing you to have to replant? What is it about trying raised beds you don't understand? Are diseases snuffing out flowers and fruits? Did you apply fungicides when you were supposed to—early, before infection started? A stitch in time ...

## KISS

Stands, of course, for Keep It Simple, Stupid. Now, good gardeners don't talk like that. It may be better to use the expression my parents used when we kids put more on our plate than we could eat: "Your eyes were bigger than your stomach." When we try to do too much—make our gardens too big or try to crowd in too many different kinds of plants—we create monsters we can't possibly take care of. Some of that tendency to overdo it comes from our perusing garden books in which we see those stunning English gardens and try to emulate them. What we forget is those gardens come with a whole corps of professional gardeners who do all of the grunt work. Grow your garden as you gain skill and time, but never let it become an all-consuming job.

## Giving begets getting

Gardeners as a class are generous, and that's the way it should be. Many plants—think iris, daylilies and hostas—will perform better if they are occasionally dug up and divided. Irises, especially, thrive after they have been taken up, broken apart and replanted, with the excess rhizomes given to friends and relatives. (Late summer is a good time of year to do that, by the way.) They will bloom all the better for you next year, and you may have created many blooming friendships along the way.

## Be bloody, bold and resolute

That's what Lady Macbeth told her husband as he expressed doubts about killing the king. I'm not encouraging regicide, but I do encourage you to be willing to prune heavily those plants that will bear better fruit and flower after a good cutting back. Grapes, peaches and plums all benefit from a bushwhacking in early spring. My dad talks about how my great-grandfather grew enormous peaches and bunches of grapes in the Germantown area of Louisville year after year after having cut the plants back to almost nothing in early spring. The cutting back also rejuvenates shrubs and many perennials. Just watch the timing. Plants that bloom in early spring should be cut back just after they bloom; plants that bloom or fruit later in the summer can be cut back when they are dormant, usually in late winter or early spring.

## IS IT DEAD?

I have gotten the following question from gardeners: "My evergreen tree turned completely brown. How long will it take to come back?" The answer is "eternity." If it's an evergreen and it's completely brown, it's dead. Just don't mistake the turning brown and shedding of leaves by the bald cypress as a sign of its demise. The cypress—a great tree, by the way—turns brown in the winter before shedding its leaves (or needles, if you prefer), which will come back in the spring. The cypress is not an evergreen.

# Garden catalogs' temptations

If you are an avid gardener as I am, you have probably spent the last several months poring over garden catalogs. You may have lusted for that perennial border with the pink peonies and blue delphiniums, coveted that rose garden on a catalog cover, and drooled over the flowering shrub border fronting that Cape Cod house.

Stop!

Before you spend your money, let's talk about gardens, catalogs and the real world.

Don't misunderstand me. I love garden catalogs. I love the lush pictures, the vivid prose, and the chance to dream about picture-perfect gardens that may be mine—someday. I have also learned catalogs can be an excellent source of information, especially when read with the eyes of experience that can sort the horticultural wheat from the chaff.

But I am also realistic. I have learned, often because of past overindulgence in catalogs, that it is best not to get carried away with what the catalogs have to offer.

I have learned that it's best, in most cases, to buy plants locally. And if you can't find them locally, you need to ask yourself the question: "Is that plant not available locally because the local nurseries have just not discovered it yet? Or have they discovered it and know it just does not do very well in our area?"

Let me give you some examples.

I was recently browsing through a catalog that offered several varieties of delphiniums. Delphiniums are eye-catching, no doubt about it, and it is hard to find a flower that offers a better true blue color than delphiniums.

If you read the catalog literature on delphiniums, you would believe they will do well for you given a little good dirt and a sunny spot.

Unless you have a greener thumb than mine, you will find delphiniums are just not going to do well for you in Kentucky. They detest our hot, humid summers. And even if they finish pouting about the heat long enough to bloom, our first spring thunderstorm will knock them silly.

But you'll never know that by reading most garden catalogs.

Or take the sourwood tree. A

gorgeous, pyramidal tree, it has white flowers in long racemes in the spring and drop-dead red fall color. Many catalogs will feature full-page pictures of this tree, it is that stunning.

But the sourwood is very temperamental about staying alive, especially if planted in heavier soils. The tree is happy clinging to a hillside in thin soil such as gardeners might have in the eastern Kentucky mountains, but if you are trying to grow it anywhere else, well, good luck.

The catalogs won't share that information with you, either.

Then there's the opposite problem.

Many catalogs feature plants you can't kill with a blowtorch, but they fail to advise you of that.

Take bamboo. You take it! And if you do, prepare to fight it for the rest of your life or until you move to a new home. Bamboo plants—most types, anyway—are extremely invasive. Some will send shoots up on the other side of an eight-foot expanse of blacktop. If you do decide to buy some bamboo, try to get your neighbor to go in on the purchase, because before long it will be on his side of the fence, too.

Finally, I would like to mention a practice I see in some garden catalogs that is deplorable. Many catalogs offer "package gardens," a combination of plants for, say, a mailbox bed or near a shady front door.

Nothing wrong with that, really, but what disturbs me are the pictures that illustrate these packages. These pictures—always drawings, which ought to tell you something—show several varieties of plants all in bloom at the same time presenting a pleasing display to the eye. Problem is, those plants in nature do not bloom at the same time.

I saw one the other day, for example, that showed white daffodils blooming in front of blue hydrangeas, a towering white-flowered clematis glistening overhead and the spikes of blue salvia next to the daffodils.

But daffodils bloom in early spring, clematis in late spring, and salvia and hydrangea in mid- to late summer.

Now an experienced gardener would know that those plants will not bloom in nature at the same time, but I am sure many novices get sucked into buying these packages and wonder what they did wrong when their garden does not look like the picture.

Those package gardens are horticultural perjury, in my opinion.

## HEIRLOOM ECSTACY

I get dozens of catalogs every winter, and one I can heartily recommend comes from Baker Creek Heirloom Seeds out of Missouri. Not only is the catalog gorgeous to look at, but the company also has an amazing collection of heirloom vegetable and flower seeds. Just beware: It will get your gardening juices flowing in the middle of January!

# Caution: orchids can be addictive

Warning: Before you read any further, know that some people have found the subject matter of this column highly addictive.

## ORCHIDS

Because of the plant's reputation for being difficult to keep alive, much less get to bloom, many a serious flower and houseplant lover will never attempt to raise orchids. Those who take up the challenge often find themselves devoting most of their time, resources and care to the reward of getting these unassuming-looking plants to turn loose of the exotic bloom within. Their lust for that spray of bloom that may come only once a year, and then with great care, is often insatiable.

Orchid enthusiasts have devoted whole rooms of their homes to their collections, built greenhouses to accommodate their passion, spent thousands of dollars acquiring rare specimens and traveled all over the country to attend orchid shows and conventions.

Teri Gimmel of Simpsonville got addicted to orchids early in life when she lived in Florida. It happened when she stepped into what she described as "the magical world of the Arcadia Sun Bulb Company." The trip was a reward for enduring the pain of tooth pulling and getting braces, and she came away with an orchid.

"Something about those tortured, twisted little plants with the gorgeous blooms appealed to me right about then," she said. "I endured a lot of pain to get into the hobby."

Gimmel's addiction went into remission while she was in college and while she taught writing at the University of Louisville, only to re-emerge full-blown when she and her husband, Rob, built a house with a studio in the backyard that served both her passion for pottery and the orchids she never completely let go of. She now tends 300 plants.

While orchids have a reputation for being among the most finicky of plants—a reputation well-deserved for some of the more than 2,000 species—Gimmel says some orchid species are no more difficult to grow than, say, African violets. She recommends the phalaenopsis, vandas and cattleyas as among the easiest to grow. As a bonus, these are also among the showiest of the orchids; they will reward the grower with a stunning spray of flowers in fall through early spring, just when their bold colors and patterns are most appreciated.

## ORCHID CARE

While some orchids are native to temperate climates and can tolerate cold (the lady's slipper, for example), think of your orchid as a tropical plant. Do not expose it to temperatures below 47 degrees. However, orchids do require a period of cool, 50 to 55 degrees, for about a week in the fall to induce blooming. Gimmel recommends putting the plants in a garage or unheated room in September or October to tell their flower buds to wake up. Just be careful not to let them get too cold.

When they are not getting the cold treatment, orchids will do well in ordinary room temperatures. They like bright, but not direct, light. A south window in summer and an east window in winter will serve. A good spot for orchids is the bathroom if it has a large window because they like humidity.

Orchids raised where the humidity is low (a typical room in the winter) should be misted occasionally, much as you would a fern.

Orchids should be watered to keep them evenly moist. They should not be allowed to dry out, but sitting in soggy soil is even worse. Watering is best done from the top of the plant. Try not to get water into the crown of the plant. Gimmel waters with a tissue at hand to wipe up any water that may get spilled into the crown and stem.

To bloom profusely, orchids need fertilizer. Gardeners can buy fertilizers formulated just for orchids—name brands Ortho, Peters, etc.—from nearly any good garden center. Read the label for fertilization recommendations. Many orchid growers make a solution of one-fourth to one-eighth of the recommended rate and give that to the orchid every time they water.

Finding orchids, especially those best for beginners, is not particularly difficult. Garden centers, the large home-supply stores such as Lowe's and Home Depot, even grocery stores, often carry the cattleya and phalaenopsis orchids. Or you can get information about orchid growing and connect with growers and suppliers by visiting www.orchidweb.org.

## QUARANTINE THEM!

Orchids and many other houseplants are prone to diseases, usually a result of overwatering, lack of sunlight or improper temperature. If your plants show signs of diseases—brown spots on leaves, yellowing leaves—keep them separate from the healthy ones. Plant diseases, like many people diseases, are contagious.

# Tool time's no joke

If I had a quarter for every time my mom used to say, "If a thing's worth doing, it's worth doing right," I would be writing this column from the beaches of Cancun. Of course, she was always talking to my brothers and sister, never to me.

A corollary for the gardener may be, "If a thing's worth doing, it's worth doing with the right tool."

A good gardening tool—I'm speaking about hand tools here—can take much of the work out of gardening. It may spare a back or a knee as well.

It is a shame how many gardeners fuss, fret and sweat with tools that are inadequate for the job. And it's a scandal how many poorly made tools are on the market, sold by every place from the priciest, high-end garden shop to the cheapest discounter. From my observations, most hand tools on the American market today are little better than worthless: blades are too thin, they are poorly connected to the shaft, the handles are crooked, or all three.

If a garden tool turns up on your Christmas shopping list this season, here's what to look for:

First, a good garden tool should feel somewhat heavy for its size, an indication it was made of heavy steel and hardwood. Some fine tool handles are made of fiberglass and they will be a little lighter than wooden tools. They have the advantage of having bright colors, a boon to one who tends to leave a shovel or rake in the weeds.

Whether wood or fiberglass, the handle should be straight. If it is wooden, check to see if the grain of the wood is straight. Also make sure the handle fits the person who will be using the tool most frequently. Take it off the rack and pretend to do on the store floor whatever you would do with the tool. You don't want a tool that will require you to bend over to work.

Look at where the blade is joined to the handle. The best tools have a blade made of a solid piece of steel that rises well up on the handle and is fastened with metal rivets through the shank and handle.

The blade should be a single piece of fairly thick steel. The hoe may be an exception to that rule; a thin-bladed hoe glides through weeds more easily than a thick-bladed one.

A good tool is not cheap, but in the long run is a better value than an inexpensive one. If you find a hoe, rake, shovel, or spade for under $15, pass it by. A good hand tool costs at least $20 to $25 and up to $40.

The British are avid gardeners, and British-made garden tools are considered the cream of the crop. One British brand is Spear and Jackson, a company that started out making war implements in the 1600s and in the last centuries has turned its swords into ploughshares–and hoes, rakes and shovels.

## Three Tools

If I were allowed but three garden tools, I would choose a trowel, a swan-neck hoe, and a garden fork.

A well-made trowel (please don't buy one with a plastic handle or blade) can be used for setting transplants and, if kept sharp along the edges, for hand weeding in tight spots.

Most hoes on the market are more useful for mixing concrete than for garden work. Their thick blades coming straight out from the handle make the gardener bend over and chop at the ground. But a swan-neck hoe is designed to allow the gardener to stand upright and pull the blade beneath the weed, severing the roots. The semi-circular blade also has two sharp points on the sides that allow for close-in weeding. Be sure to match the handle to your height.

A well-made garden fork will dig as well as a shovel or spade in our clayey soils while at the same time loosening and breaking up the soil as you dig. Most garden forks are sold with short handles, but a long-handled one will give you more leverage, especially useful if you have rocky soil or dig-and-divide perennials such as ornamental grasses and daylilies.

If you can't find a good tool in a nearby hardware store or garden center, one excellent mail order source is Gardener's Supply, 128 Intervale Road, Burlington, VT 05401; www.gardeners.com

## COOL TOOL

Some of the sorriest garden tools on the market are pruners. It's easy to find pruners that dull quickly, fall apart easily or chew up the stem rather than cut it. That's too bad because pruning is an essential chore, and one made miserable by a poorly made tool. I strongly recommend you pay the extra few bucks and buy a pruner made by Felco. I like the bypass pruner rather than the anvil type, but whichever you choose, get a Felco!

# Gardens aren't just for plants

Browse a garden center or thumb through a garden catalog and you'll notice considerable space dedicated to garden art objects.

Gardens are not just about plants these days.

It's true, gardeners have always "decorated" their gardens with objects other than plants.

Remember tractor tires painted white, made into a crown, and planted with petunias or marigolds?

Remember bathtubs set up in the garden, painted blue on the inside and surrounded with rose bushes?

Remember the little white hens made of concrete with little yellow chicks following in their wake under the taxus bushes?

If you wanted to make a high-culture garden statement, you put a handkerchief on your white concrete goose.

Remember pink flamingos?

Mercifully, garden art has come a long way.

Gardeners can choose from a wide range of items that complement the beauty of the plants. Many of these garden art objects also serve useful purposes as well.

One object d'art that is enjoying a revival is the gazing globe.

Gazing globes were originally made of glass, set upon a pedestal and placed amid the formal or cottage garden. The reflection of the blooms in the glass provided a two-for-one visual treat.

Most globes made these days are fashioned of polished steel, making them less vulnerable to destruction by an errant baseball or a long-tailed Labrador, but their reflective surface is just as fetching. Gazing globes are available that float on the water; they shimmer among the water lilies like giant soap bubbles.

Garden centers are also offering more statuary, everything from a towering Neptune who, with his trident, stands watch over a bog garden, to a diminutive cricket, frog or rabbit that adds visual interest to a small corner of a flower bed.

One company offers a catalog full of dozens of gargoyles designed to decorate the garden and to chase away the evil spirits as they were meant to do from their perches on medieval cathedrals.

While gardeners spend a great deal of time shooing away bugs from their plants, many are adding them to their beds—copper, brass and wrought iron insects.

## ART WITH A PURPOSE

Garden art isn't merely decorative.

Sprinklers, sculpted of copper, add a classy touch and water the plants when needed.

As gardens get smaller, gardeners are paying more attention to plants that grow vertically. Wrought-iron or redwood arches or trellises can gracefully support a flowering clematis or climbing rose and add a decorative touch to the garden even in winter.

Bamboo is making a comeback as a material that is both useful and beautiful in the garden. Bamboo teepees can hold up pole beans as they have for generations, but bamboo is also used to make trellises for climbing roses, special structures to protect plants from marauding animals such as deer, and to build fences for Japanese-style gardens.

Perhaps one of the cleverest garden art objects I have seen is a glazed ceramic snail ornament that serves as a slug trap.

I have never considered it my job to tell adult gardeners what is tasteful and what is not. I do not intend to start now by offering advice about selecting and placing garden art.

I might humbly suggest, however, that the gardener keep in mind when choosing garden art that the best-looking art objects complement and do not overwhelm the plants. In other words, a gargoyle may look great in the garden. A gaggle of gargoyles gags.

Otherwise, anything goes as far as I am concerned.

Even pink flamingos.

## SCOUR THE TRAIN TRACKS

Those glass insulators they used to put on electric poles make great little garden decorations. They capture sunlight and sparkle with color reflected from your flowers. The blue ones are especially nice sitting among some yellow or orange marigolds. You can buy them at antique stores or often find them lying along railroad tracks. Of course, you're trespassing when you walk along railroad tracks. I won't tell you where I got mine.

# A plant by any other name ...

A friend once asked me to come diagnose some problems in her yard.

Her buttercups were not blooming.

Buttercups? I thought. She has buttercups?

Then it dawned on me what she was referring to. What she was calling buttercups was what I, and what I thought the rest of the world, called daffodils.

It dawned on me because I remembered once Byron Crawford admitted in a column that he also calls daffodils buttercups. Of course I did not tell my friend that. Instead, I told her I thought everyone who called daffodils buttercups had been dead for 20 years.

And I also told her not to worry; for some reason daffodils just did not bloom well that year all over the area.

Then another friend asked me to come take a picture of her snowball bush.

"It's the most beautiful I have ever seen it," she said.

But I was thinking there is no way she had a snowball bush blooming. Hydrangeas, which is what I call snowball bushes, do not bloom until late June. She was right. Her snowball bush was beautiful, but it was viburnum, not hydrangea.

These examples illustrate what happens when people communicate about plants using common names. The name for a plant in one area can be the name for a different plant in another. Plant names even vary from one person to the next.

You say toMAYto, I say toMAHto. You say snowball bush, I think hydrangea, and so on.

The confusion created over common names of plants is why botanists assign Latin names to them and why avid gardeners should learn a little botanical Latin. It's not too difficult to pick up enough botanical Latin to make sure you put the right plant in the right place or pick out the plant you intended at the garden center. The best way to do that is to start paying attention to Latin names when you see them in catalogs or on plant tags.

All plants have a Latin genus and species name. For example, the botanical name for a white oak is Quercus alba. Quercus refers to the genus oak; alba refers to a particular kind of oak, in this case white, as opposed to a pin oak or a black oak.

If the plant comes in a cultivated form, the variety name will follow the species name. For example, a commonly planted shrub (and a nice plant for shade, I might add) is Hydrangea quercifolia 'Snow Queen.' So what you have there is an oak-leaved hydrangea, variety Snow Queen. See the Latin words for oak and foliage in the name quercifolia?

## The Right Plant

Am I saying you are less of a gardener if you do not know botanical Latin? What I am saying is that having a smattering of botanical Latin at your fingertips may keep you from making some mistakes in how you treat your plants and in what plants you select for your yard.

For example, the same friend who called daffodils buttercups asked me if I would come cut down a magnolia tree for her because it had lost all of its leaves.

What she called magnolia is probably the same tree most people picture when they hear "magnolia." It is Magnolia grandiflora, a symbol of the deep South, and also a tree that keeps its leaves all year. And if what she had was Magnolia grandiflora, she would be right—the tree was probably dead if it had lost its leaves.

But what she had was a deciduous magnolia, probably magnolia soulangeana, which is supposed to lose its leaves in the winter. The tree was not dead; it was just waiting for warmer weather to put out leaves.

Here's another example: Suppose you are in a garden center and you see a plant labeled "phlox." If all you know about that plant is that it is a phlox, you don't know much. There is Phlox subulata, which is commonly called creeping phlox; Phlox stolonifera, commonly called woodland phlox; and Phlox paniculata, commonly called summer phlox. And that's just a few of the many "phloxes."

If you want the spring-blooming groundcover that looks good planted among rocks or hanging over rock walls, you want subulata (sub = below – get it?) If you have loose, woodsy soil and want a taller spring bloomer than subulata, you want stolonifera. And if you want a plant that blooms in the summer, you want paniculata, which will bloom throughout the summer and into the early fall, long after the other two have quit blooming.

So in the plant kingdom, the language of the Caesars is still alive and well. And a rose by any other name may indeed smell as sweet, but are you talking about Rosa rugosa, Rosa davidii or Rosa pisocarpa?

There is a difference.

## A Vulgarity

If you pay attention to the Latin names of many plants, you may see the term "vulgaris," as in Phaseolus vulgaris (green beans) or Syringa vulgaris (lilacs). Those plants are not garden porn and will not do scandalous things. The Latin "vulgaris" means "common." So plant without worry of offending the neighbors.

# Gardeners will save the world

It's getting hotter. Well, that's what the scientists and meteorologists tell us. And they say it's going to keep getting hotter because we're spewing pollutants into the air and replacing our forests and fields with bricks, asphalt and cement.

I do wonder how the same people who can't accurately predict the weather next week can tell us what the climate will be like 10 years from now. Still, the evidence—polar bears skating on thin ice and the snows of Kilimanjaro melting—is suggesting a hotter, if not hellish, future for all of us. They call it global warming.

And depending upon whom you listen to, that will mean all sorts of dire consequences. The Midwest will become a desert. Coastlines will be flooded and Florida will shrink to about 10 miles across. Hollywood will be underwater (there are always some positives in every disaster).

We can wait for the government, or world governments, to do something about global climate change, but that would be unAmerican. We, especially we gardeners, need to find solutions and let the government follow or get out of the way.

What can we urban, suburban and rural gardeners do? Plenty. Have you ever driven from the city to the countryside on a summer day with the car windows rolled down? Did you notice that when you left the concrete jungle the air cooled as the landscape turned greener? That's right. It's cooler in the countryside. Why? More plants. More trees. More grass. And less concrete and blacktop to trap the heat.

So gardeners can help cool down at least their own little part of the world by adding more plants. Here are a few suggestions:

- Plant trees. Trees are the planet's lungs. They filter out the gunk in the air and cool the surrounding air. Every gardener should plant at least one tree in the spring and one in the fall. If you have the room, plant the large shade trees that are native to the eastern U.S. Oak, ash, hickory, hackberry, maple, bald cypress, sweetgum, catalpa, and Kentucky coffee tree are all good choices. Remember to look up, down and around before you plant a large tree. Don't plant large trees where they will grow into your roof, get tangled in power lines or send roots into sewer lines. Even city gardeners with a small lot have space for small trees. Serviceberry, dogwood, hawthorne, redbud and fringetree will fit almost anywhere.

- Lose the lawn. Or as much of it as possible. Grass is, in fact, a good air cleaner and helps keep the land cooler, certainly cooler than concrete. But we in the United States do not just have grass. We have grass carpets that have to be constantly doused with polluting fertilizers and chemicals. Then we roar over them once a week with lawnmowers that, pound for pound, produce about 20 times the pollution automobiles send into the air. Try gradually replacing lawn with mulched beds of shrubs, trees and perennials.

- Grow your own food. Or as much as you can. And patronize those growers and farmers who live near you as much as possible. Each calorie of food we consume in this country represents several calories in energy we spend getting it to the table because so much of what we eat is shipped in from someplace else. The energy burned bringing us food enhances the greenhouse effect that is making the earth warmer. Even if all you have is a balcony outside an apartment, grow some of your own food. Grow tomatoes and greens in tubs. Tear out some of the lawn and grow beans on tepees, cucumbers up a fence, tomatoes in cages and squash in beds, and you will be surprised how much food you can produce in a small space.

- Plant shrubs. If space is tight, shrubs can provide shade and homes for wildlife where trees would be impractical. Viburnum, calycanthus, clethra, buttonbush and elderberry are excellent native shrubs.

- Plant tough perennials and annuals. If your garden is full of sickly plants that require constant watering, fertilizing and spraying for insects and disease, get rid of them. Some tough perennials include salvia, sedum, nepeta (catnip), peonies, shrub roses, rudbeckia, true geraniums and ornamental grasses. Tough annuals include zinnias, marigolds, amaranth, and impatiens (in shade).

Here's the point. If the hotter-than-fire forecasts do come true, gardeners may save the world. If the forecasts are wrong, we've still done a wonderful thing when we've covered up our space with plants.

## WORMS, WORMS EVERYWHERE

In the spring, after a heavy rain, my driveway is covered with hundreds of red worms. That is a good thing. A soil full of worms is a healthy soil. Worms aerate the soil, provide their own fertilizer and open up passageways for roots. Never apply an insecticide that is harmful to earthworms.

# Five garden questions

In the garden, experience is always the best teacher. But going into gardening armed with a certain basic knowledge isn't a bad idea either.

Below are five questions and answers. If you can answer those questions for your own garden situation, you may save yourself tons of money and years of heartache.

## Q: What is the dirt on your soil?

A: Knowing your soil type is the key to knowing what will thrive and what will fizzle in your garden. There are three basic soil types and hundreds of blends in between. The three types are clay, sand and loam. Clay soils are heavy, usually fertile and slow to drain. Sandy soils are often infertile, but light and well drained. Loams are something of a combination of the two—heavier than sand, better draining than clay and, on the whole, the best type of the three.

Most gardeners in Kentucky have a clay soil. Eastern and southern Kentucky gardeners are more likely to have a sandy soil. Some western Kentucky gardeners have a loam. Your county Extension agent can tell you what type you have if you can't tell by pulling up a shovel full.

In small spaces, soil type can be improved by adding organic matter, such as compost or animal manures. But in larger areas, it's better to find the plants that fit the soil rather than make the soil fit the plant.

## Q: Acid or alkaline?

A: Knowing the soil's pH is the second key to successful gardening. Some plants will not tolerate an acid soil while others, such as azaleas and potatoes, will thrive in it. Soil pH is measured on a scale from 1 to 14, with 1 being extremely acidic and 14 being extremely alkaline. Most of the state has soils that are nearly neutral, which is good because that suits most plants. The soils of mountainous eastern Kentucky are the most likely to be acidic.

Like soil type, pH can be manipulated a small scale by using soil amendments. Sulfur makes soil more acidic; lime makes soil more alkaline. But before you pour out the cash for amendments, take a soil sample to your Extension Service office and ask for an analysis. Then consider getting plants that like the soil pH you have rather than trying to play tricks with Mother Nature.

## Q: How big is that plant?

**A:** It is surprising how many gardeners will buy plants in a store without a clue to their ultimate size. Knowing a plant's dimensions is critical for practical as well as aesthetic reasons. Planting a tree too tall for its space, for example, leads to its mutilation somewhere down the road because it climbed up over a roof or into utility wires. Planting a tree is an act of generosity toward future generations, but not if they will have to pay a tree trimmer to undo your work 80 years from now.

Aesthetically, plant placement in the border or flowerbed depends upon size. A small plant in the back of the border is lost to the eye. A large plant in front hides the rest of the bed. Look at the plant tag. Or if the plant does not have a tag, get the plant's Latin name and variety and look it up in one of the resources at right. And if the plant doesn't have a tag, look somewhere else.

## Q: Can you name your poison?

**A:** Herbicide use in lawns and gardens is on the rise because surveys show that weeding is gardeners' most common and most despised job. Yet many gardeners do not understand herbicides and how they work. That lack of knowledge can lead to disaster.

Herbicides come in three basic types: total kill, selective kill and pre-emergent.

Total kill herbicides—Roundup is the most common—will kill all existing plants, which is desirable when you are

## GARDEN RESOURCES FOR KENTUCKY GARDENERS

I find myself using the following books repeatedly when I come across something in the garden I haven't seen before or when I need a little inspiration:

*Insects and Gardens* by Eric Grissell

*The Southern Living Garden Book*, 2004 Edition

*The Southern Gardener's Book of Lists* by Lois Trigg Chaplin

*Dirr's Hardy Trees and Shrubs* by Michael Dirr

*The Organic Gardener's Handbook of Natural Insect and Disease Control* edited by Barbara Ellis and Marshall Bradley

*American Horticultural Society's Encyclopedia of Perennials*

going to renovate a lawn and want to start all over. It is not desirable if you want to take out the weeds but leave the desirable plants. Spray carefully, or do it the old-fashioned way—hoe! Seed can be planted after spraying with a total kill herbicide because it does not harm what is in the ground, just what is green and growing.

Select kill herbicides eliminate a certain type of plant but not others. Herbicides containing 24-D, for example, kill broadleaf plants but not grasses. Again, useful in the lawn but not so good anywhere else.

Pre-emergent herbicides are probably the least understood. They are sprayed or spread on the ground to kill seeds lying dormant. Used correctly, they will not kill what is already growing. Preen is one brand name. Pre-emergents are used in flower and vegetable gardens to keep weeds from sprouting. They are also used to kill crabgrass if spread at the proper time. Always read the label carefully and follow it. Pre-emergents, if spread too heavily, can kill existing plants.

**Q: What's bugging you?**

**A:** Fortunately, many gardeners are learning that not every bug in flower and vegetable gardens is the enemy, and they are avoiding insecticides that kill indiscriminately. Knowing what bug is doing damage and how best to control it saves both time and money and holding off on the spraying is better for the environment. Many bugs do little damage if left uncontrolled, and many of the bad guys can be controlled without resorting to poisons. Knowledge is the key.

## SUNFLOWER DESERT

Bird lovers who feed wild birds sunflowers from feeders often notice that the soil beneath the feeder becomes bare. What is happening is that the sunflower seeds release a chemical that kills vegetation—nature's way of taking out the competition for a germinating sunflower seed. It's called an alleopathic reaction. To avoid bare patches beneath the feeder, try moving it every couple of weeks or so.

# HELP!
# Squelch invasive plants

They're creeping, crawling, climbing and crushing native vegetation. They're spreading by seeds, rhizomes, stolons and spores. They're invasive plants, and they're a problem.

Kentucky, like the rest of the nation, is at risk of being swallowed by invasive plants.

Invasive plants are non-native species that threaten to displace native wild vegetation and, in some cases, have already driven native plants to extinction. Invasive species can disrupt natural bird, insect and mammal ecology. Because they come from other places, exotic species often thrive at the expense of natives because they do not have in this country the enemies that controlled them in their homeland.

Some invasive plants have come here accidentally, but most have come here deliberately. Some were brought with good intentions that had bad results. Kudzu, for example, which thrives as far north as Louisville, was brought by the U.S. Department of Agriculture from Japan to control soil erosion. It has swallowed much of the South.

Agriculture officials once recommended the planting of multiflora rose to offer habitat for wildlife. It has become a bane of farmers and often smothers out more beneficial plants. For years, highway officials planted crown vetch to control soil erosion on roadcuts. It does that. It also swallows up native vegetation.

What does this have to do with gardening in Kentucky?

First, gardeners are responsible for unleashing many of the invasive plants into the landscape. Purple loosestrife is commonly sold in nurseries and garden centers, where gardeners pick it out for is handsome purple flower stalks that bloom throughout the summer. But purple loosestrife has escaped into the wild, where it has smothered native vegetation, especially along ditch lines and other wet habitats that it favors. It is a severe problem in the upper Midwest and is now found in 21 counties in Kentucky, most in the central part of the state.

Two other severely invasive specimens favored by gardeners are wintercreeper and burning bush. Wintercreeper is bred in dozens of cultivars that gardeners use as small shrubs and groundcovers. The variegated varieties are especially attractive. But wintercreeper, in places, swallows native vegetation, including large trees. A section of Clear Creek Park in Shelby County has large hackberry, black cherry and walnut trees smothered by wintercreeper. Gardeners should also know that wintercreeper, especially the dark green, climbing variety, is prone to powdery mildew and really is not a pretty sight. Burning bush is spectacular in the fall with its glowing red leaves. But it has escaped into the wild where it has displaced native shrubs.

In fact, many of the plant species local gardeners and farmers favor are invasive and a threat to the wild: lespedeza, Ky. 31 fescue, Japanese honeysuckle, Japanese barberry, and English ivy. Even our beloved bluegrass is an exotic species that displaces native grasses.

### A CALL TO ACTION

The Kentucky Exotic Plant Council, based in Lexington, was formed in 2000 to call attention to the problem of exotic and invasive species threatening to overwhelm the natives. The council lists 93 exotic and invasive plants in three categories:

SEVERE THREAT to native plants, including purple loosestrife, garlic mustard, tree of heaven, crown vetch, burning bush and multiflora rose.

SIGNIFICANT THREAT to native plants, including mimosa, Japanese barberry, empress tree and white mulberry.

LESSER THREAT to native plants, including daylily, lambs quarter, catnip and St. John's wort.

Notice that many of these plants are

## THE DIRTY HALF-DOZEN

Every year, the Kentucky Exotic Plant Council names a "Most Wanted" invasive plant. Most wanted, in this case, means least wanted. Here are the six listed since 2001 and native plant alternatives:

**Purple loosestrife**—One of the most damaging, chokes out native wetland vegetation. Alternatives are great blue lobelia, obedient plant and blazing star.

**Wintercreeper**—spread by birds and lateral shoots, this groundcover can damage trees. Alternatives are ginger, Allegheny spurge, mountain lover or cliff green.

**Crown vetch**—Spread by seeds and creeping root system, it smothers native grasses and shrubs. Alternatives are a prairie mix, including dropseed, goldenrod, blazing star and bluestem.

**Burning bush**—A popular landscape shrub, seeds are spread into the wild by birds. Alternatives are strawberry bush, spice bush and winterberry holly.

**Chinese grass**—Another popular landscape plant that can escape and outcompete native grasses. Alternatives are switchgrass (panicum), and Indian grass.

**Asian bittersweet**—An attractive but highly invasive vine. Alternatives are trumpet honeysuckle and American bittersweet.

favored by gardeners and farmers alike. And nothing is going to keep gardeners and farmers from planting these species. Fescue is a staple of hay fields, and daylilies are a joy in the summer garden.

But one thing gardeners can do to help protect native species is plant the invasive and exotic species in places where they are not likely to escape into the wild. Many invasive plants spread by seed, for example. They can try to keep those plants from going to seed. Gardeners can also make sure invasive species are not planted near wild areas where they are likely to migrate. For example, a horticulturist friend of mine knows about the threat of purple loosestrife, but he has some in his own garden. But the plant is in a bed surrounded by lawn, and has no chance of escaping into a stream or ditch.

The other action gardeners can take is to buy native alternatives to invasive plants. The Kentucky Exotic Plant Council has a list of native plants that have similar habits to the exotic species. More and more local nurseries are starting to stock these natives—plants you can feel good about putting into the ground.

## A WOODLAND BULLY

Bush honeysuckle is horribly invasive and crowding out more and more of Kentucky's native shrubs and vines. Birds love the red seeds and spread the invader with their droppings. You'll find bush honeysuckle coming up in your flower beds, among your evergreens and shrubs—just about anywhere the birds have dined on the seeds. Just cutting it back will not take it out. You have to cut it down, wait for it to start to emerge again, then spray with a weed or brush killer. It may take several applications of the herbicide to finally kill off bush honeysuckle.

# Are you mooning?

I have been gardening since I was barely old enough to take up a hoe and, with my sister in tow, follow my uncle "Red" to the tomato patch, where we proceeded to chop down the tomato plants and cultivate around the weeds.

I'd like to think my technique has improved since those tender years, but even as I grew as a gardener, so to speak, I occasionally would get lectured by the "old-timers" about why and how to cultivate my garden by the moon signs. An avid reader of the *The Old Farmer's Almanac*, I vaguely understood the concept of moon sign gardening, but also knew that I didn't have time for any restrictions on when and what I planted. Gardening was, and still is, something I do in an otherwise busy schedule, and I didn't have time to figure out whether the signs were right for planting tomatoes, pulling weeds or transplanting onions. Those things got done when I had the time; the moon was in the sky, and I was down here on the ground doing my gardening when I got home from work, the kids were napping and the animals fed and watered.

But now that I might qualify as an old-timer myself—one of the first magazines I wrote for having appeared in an antique store, after all—I thought it would be fun to offer information about moon sign gardening. I am sure there is a whole generation of gardeners who do not have a clue about gardening by moon signs and who may want to try it, if only on an experimental basis. Use the information as you will—or not at all—but know that many gardeners swear by the advantages of planting, transplanting, weeding, spraying and harvesting during the proper phases of the moon.

## Light of the Moon

Gardening by moon signs is based on the observation that the phases of the moon affect the earth. Anyone who lives along the coast knows the moon is responsible for tides. And for years scientists have documented the effects of the moon on people's moods. Emergency rooms, psych wards and the police are all busier during a full moon, as the pull of that heavenly body reaches its zenith.

Gardeners who use moon signs as a

guide to planting times believe the moon has an effect on plants, too. And there is some scientific evidence to support their beliefs, as well as plenty of evidence that does not.

Basically, the waxing moon (getting larger—from new moon to full moon) is the period of growth above ground, while the waning moon (getting smaller—from full moon to new) is the best time for growth below ground and for getting rid of undesirable elements in the garden, such as pests and weeds. So in a waxing moon, you would plant flowers; vegetables, such as corn, tomatoes, beans and squash; and flowering trees, shrubs, lawns and fields where you want vigorous growth. In the waning moon, you would plant vegetables that grow underground, such as potatoes, beets and carrots; transplant trees and other large plants where strong root growth is desirable; spray harmful insects; and pull weeds.

So if you're anxious to get started in the garden and want to try planting by the moon signs, follow this example. If the full moon occurs on March 1, that means between March 1 and March 15, the moon is waning, and it's a good time to plant underground vegetable crops. If you want early peas, plant them before March 1 or wait until the moon is new again, on the 15th. From March 15 until March 30 is a good time to plant vegetable crops where the above-ground portion is desirable, such as peas, spinach, broccoli and cabbage. Early flowers, like pansies and creeping phlox, also should be planted between March 15 and March 30 if you follow moon signs. This is only an example. Please consult a *Farmer's Almanac* or another source to determine the dates of the moon's phases.

## HAVING FUN YET?

Realistically, what any gardener who wants to plant by moon signs without earning a degree in astronomy should do is get a chart or table, typically published in sources like *The Old Farmer's Almanac*, and plant on the best planting dates provided.

If you're like me when I tried to experiment with moon-sign gardening, I found that the two or three best planting dates every month would always correspond to a period of heavy rain or someone's funeral or wedding. So the garden got planted when the garden got planted and it grew and flourished, most of the time, but who knows what I could have achieved if I had just "planted by the signs."

### COMPLICATIONS

Now, of course, as any good gardener knows, just because the sign is favorable, doesn't mean you run out and plant anything that fits the waxing or waning moon. You've got to watch for frost dates, soil moisture and a number of other variables.

Easy, right? Well, it can get much more complicated. Many moon-sign gardeners also plant according to the zodiac, which muddles things immensely. The 12 signs of the zodiac are divided into the elements of air, fire, water and earth and are barren or fruitful depending upon their element. Pisces, for example, is a very productive, moist sign, while Aquarius is an air sign, barren and dry. Approximately every two-and-a-half days the moon moves into a different sign. So what that means is the absolute best days for planting above-ground crops are times of a waxing moon that is moving through the productive signs of the zodiac—Pisces, Capricorn, Libra, Cancer, Scorpio and Taurus. On the other hand, if you want to cut grass to slow growth, harvest, kill weeds or spray bugs, do it when the waning moon moves through the barren signs: Aries, Gemini, Leo, Virgo, Sagittarius and Aquarius.

## MOONFLOWER

Like to sit outside at night? Why not plant a moon garden close by your favorite chair? Moon gardens typically feature all-white flowers, some heavily scented. A great favorite for the moon garden is moonflower (Ipomoea alba). It is a climbing vine that produces a fragrant white flower that blooms at night. Moths love it, and you will too.

# About the Author

Walt was born and raised in the south end of Louisville, Ky. He has a degree in English from the University of Louisville and a master's degree in journalism from Indiana University. He taught high school English and journalism for three years, and then taught at the University of Louisville and Wright State University in Ohio for more than 20 years. While teaching, Walt was also a freelance writer for many regional and national publications; he specialized in farm and garden subjects. He wrote a garden column for the original *Kentucky Monthly* magazine in the late 1970s and early 1980s. He also wrote a garden column for *Dayton* magazine in the early 1980s. From 2000 to 2009, Walt was reporter, photographer and editor of the *Sentinel-News* in Shelbyville, Ky. He started writing Kentucky Gardening, his column in *Kentucky Monthly* magazine, in 1999. Currently, Walt is horticulture technician with the Shelby County Cooperative Extension Service. He lives in Simpsonville with his wife, Mary Lou, where he operates a small farm.

# Index

## A

achillea (see also yarrow) ...26, 52, 83, 118, 131
aconitum .................................................69, 83
acorn squash ...............................................98
All-America Selections award ....................38
allium....................................................... 111
aluminum sulfate........................................22
alyssum ............................................... 86, 97
amaranth ................................................. 207
amaranthus .............................................. 186
American beech .......................................163
American boxwood .................................. 180
Amish Deertongue lettuce........................ 190
amur maple ..............................................183
*Animal, Vegetable, Miracle: A Year of Food Life* ..... 30
annual sage................................................24
apples.........................................................74
arborvitae.................................................172
artemesia....................................................68
asclepias ................................... 117, 118, 131
ash tree ............................................ 175, 206
  mountain ash tree...............................185
Asian bittersweet ..................................... 212
Asiatic lily................................................ 160
aster..........................................................185
astilbe ....................................... 63, 69, 80, 83
autumn clematis....................................... 124
azaleas...........................20-22, 51, 194, 208
  flame azalea ..........................................22
  sweet azalea ..........................................22

## B

baby's breath..............................................68
bachelor button ....................................... 186
  Blue Boy bachelor button ................... 190
bacterial wilt ............................................101
Baker Creek Heirloom Seeds......... 38, 137, 197
bald cypress ....................47, 120, 163, 195, 206
bamboo .................103, 104, 105, 174, 197, 203
banana squash ...........................................98
bark................................................... 181-183
basil ...........................................................93
  purple basil ...........................................92
beans...............................................30, 59, 60
beautyberry .............................. 47, 139, 140
bee balm.................................................. 190
beech ............................................181, 182, 183
  American beech ............................ 163, 181
begonia ......................................................50
Bella Donna rose...................................... 190
bergenia .....................................................83
berm................................................ 50, 119, 174

Bernheim Arboretum and Research Forest
 ....................................................... 170-171, 184
Bernheim Select......................................... 171
berries.................................................139-141
  blackberries..........................................185
  blueberries ........................................... 194
  cherries .................................74, 75, 76, 77
  raspberries........................................... 148
  strawberries .............................. 64-66, 148
birch trees .............................. 56, 182, 185
  dwarf river birch...................................83
  river birch.................................... 50, 120
  white birch ..........................................181
birds................................................... 184-186
bittersweet ............................................... 186
  Asian bittersweet................................. 212
black-eyed Susans (see also rudbeckia)
 ...............................................................180, 185
blackberries..............................................185
blackberry lily ............................................48
blackspot.........................................19, 33, 36
blood meal .................................................72
Blue Boy bachelor button ......................... 190
blue flag iris ....................................... 54, 55
blue spruce...............................................172
bluestem grass.................................... 111, 185
blueberries ............................................... 194
bluegrass ........................................... 164-165
boulevard cypress .................................... 180
boxwood .................................56, 91, 104
  American boxwood.............................. 180
Brandywine tomato ..39, 123, 127, 128, 129, 190
broccoli ............. 13, 14, 15, 60, 85, 92, 100, 215
brown-eyed Susans...................................185
Brussels sprouts .............................. 13, 15, 85
Bt ........................................... 86, 101, 116
bud union.............................................35, 36
buddleia (see also butterfly bush). 117, 118, 131
bulbous oatgrass.......................................185
bulbs ..............................................134, 142, 149
bur oak ............................................. 162-163
burning bush .............................. 180, 211, 212
bush honeysuckle .....................................213
buttercup .................................................204
butterflies................................................ 116-118
butterfly bush (see also buddleia)
 ...........................................................24, 89, 116, 131
butternut squash .......................................98
buttonbush...............................................207

## C

cabbage ............................... 13, 14, 15, 85, 215
  Chinese cabbage ................................... 13

cabbage maggot ............................................14
cabbageworm .......................................14, 15
California poppy............................................186
calycanthus ................................120, 147, 207
Canadian ginger ............................................80
canna ...............................................................120
cantaloupe.................................... 101, 135-137
Captain Jack's Deadbug Brew Flower and
 Vegetable Garden Dust ..........................39
cardinal climber ...........................................190
cardinal flower ..............................................120
Carloftis, Jon ........................................ 179-180
carrots............................................. 85, 146, 215
catalpa.................................................... 163, 206
catmint (see also catnip, nepeta) ....23, 24, 69
catnip (see also catmint, nepeta) ....... 118, 212
cauliflower ....................... 13, 14, 15, 29, 30, 85
celosia ...............................................................92
Cercis chinensis..............................................45
Cercis texensis ...............................................44
Charles De Mills rose...................................190
chaste tree .....................................................180
cherries ..............................................74, , 75, 76, 77
cherry tomato .....................................92, 93, 128
 Large Red cherry tomato .....................190
cherry tree .....................................................185
chestnut blight ..............................................177
Chinese cabbage ............................................13
Chinese dogwood.........................................185
Chinese grass ................................................212
chives ........................................................92, 93
chokeberry ....................................................147
chrysanthemum..............................................68
cinnamon fern ..........................................62, 80
clary sage (see also Salvia sclarea) ........... 107
clay ..................................................................208
clematis.......... 23, 24, 47, 112-113, 126, 197, 203
 autumn clematis ....................................124
 clematis Jackmanii............. 24, 112, 113, 124
 sweet autumn clematis .....................47, 112
cleome............................................................. 180
clethra................................63, 91, 120, 147, 207
climbing hydrangea ..............................124, 126
climbing rose.........................................124, 125
climbing wisteria..........................................125
clover ................................................................87
cockscomb.....................................................190
colchicum.........................................................83
cole crops ..................................................13-15
coleus .......................................................24, 180
collards ............................................................13
columbine ........................................ 24, 51, 152
coneflower (see also echinacea). 24, 26, 47, 48,
 57, 103, 104, 105, 118, 147, 152, 180, 185, 186
coreopsis ........................ 51-52, 110, 85, 88, 185
corn..................................................60, 70-73, 101, 135
 sweet corn ....................... 70, 72, 73, 100, 135

corn borers ......................................................73
cosmos ..................................................48, 69, 180
cottage gardening .............................. 189-191
cottonseed meal ............................................72
crabapple ......................... 83, 117, 140, 141, 185
crape myrtle .................................................. 182
crocus ................................... 23, 142, 145, 149
crossvine ..........................................................47
crown vetch ........................................... 211, 212
cucumber beetle ..................... 60, 99, 101, 159
cucumber magnolia .....................157, 163, 183
cucumbers...............................58, 59, 60, 92, 93, 101, 207
cucurbits .........................................................101
cutworm......................................................... 15
cypress
 bald cypress.................47, 120, 163, 195, 206
 boulevard cypress .................................. 180
 dwarf hinoki cypress ............................. 180
 Leyland cypress .......................................174

# D

daffodil .62, 142, 144, 145, 149-150, 190, 197, 204
 thalia daffodil ...........................................57
dahlia .............................................................. 180
daisies
 gloriosa daisy........................................... 186
 Shasta daisy...................................... 69, 152
daphne ........................................................... 190
dawn redwood .............................................163
day-neutral strawberries.............................65
daylily ...... 48, 49, 56, 67, 110, 115, 117, 124, 180,
 195, 201, 212, 213
deciduous holly ............................................ 140
delphinium .............................................51, 190, 196
dethatch ......................................................... 148
digitalis .............................................................83
dill .....................................................................87
Dixie Queen watermelon ............................39
dogwoods ............43, 44, 115, 117, 180, 185, 207
 Chinese dogwood....................................185
 kousa dogwood ...............................82, 140
 red-osier dogwood .......................120, 185
dusty Miller.................................................... 180
Dutch elm disease .......................................177
Dutch iris ................................................ 23, 145
dwarf conifers ................................................91
dwarf fruit trees .............................................75
dwarf hinoki cypress .................................. 180
dwarf itea ........................................................91
dwarf river birch ...........................................83

# E

E. coli contaminations ..................................27
Earth Friendly Expanding Potting Mix ......39
earworms .........................................................73
Eastern hemlock ..........................................185
Eastern redbud............................................. 180

echinacea (see also coneflower) .................47
echinops.......................................... 23, 24, 111
eggplant ..................................................86
    Listada De Gandia eggplant .................190
elderberry ................................ 147, 185, 207
emerald ash borer....................... 102, 175, 176
empress tree ......................................... 212
English ivy................................104, 186, 212
English lavender .................................... 190
epimedium ............................................. 111
Epsom salts ..............................................36
euphorbia ............................................... 111
ever-bearing strawberries ..........................65

# F

false indigo............................................. 111
Felco ..................................................... 201
ferns............................................ 62, 63, 78-80
    cinnamon fern ................................ 62, 80
    Japanese fern ........................................80
    painted fern ..........................................63
fescue...............................................212, 213
    Ky. 31 fescue...................................... 212
Festiva Maxima peony ............................. 190
feverfew .................................................. 87
firethorn (see also pyracantha) ..........140, 174
fiveleaf akebia......................................... 186
flame azalea.............................................22
*Flora: A Gardener's Encyclopedia* .....................187
flowering quince......................................47
flowering tobacco ............................ 180, 190
foamflower ..............................................63
forsythia................................................. 190
fothergilla...........................................63, 91
foxglove................................................. 190
    Giant spotted foxglove........................ 190
Franklin tree...........................................183
French marigold ..................................... 180
fringetree............................................... 207
fungicide ............................... 85, 86, 87, 194
fuzzy deutzia ...........................................47

# G

gaillardias................................................38
gamma grass .......................................... 120
garden art............................................ 202-203
garden islands ..................................... 81-83
garden tools ..................................... 200-201
garlic ..................................................... 146
garlic mustard ........................................ 212
gentiana ..................................................83
geranium .......................................... 50, 83
    hardy geranium ................................. 111
    true geranium .................................... 207
German bearded iris. 23, 26, 53, 54, 55, 56, 190
Giant spotted foxglove ............................ 190
ginger.................................................... 212
    Canadian ginger ....................................80
ginkgo........................................... 46-47, 162
gladiolus ..................................................56
global warming.......................................206
globe thistle ....................................152, 185
gloriosa daisy ......................................... 186
goatsbeard....................................... 62, 80
golden rain tree ........................................47
goldenrod......................................... 48, 185
grape hyacinth.................. 23, 57, 142, 190
grapes................................................... 195
grasses...............................................185, 201
    bluestem grass ...............................185, 212
    bluegrass ............................. 164-165, 212
    bulbous oatgrass .................................185
    Chinese grass..................................... 212
    gamma grass ..................................... 120
    maidenhair grass ............................... 180
    ornamental grass ............ 147, 172, 201, 207
    ribbon grass.........................................69
    rushes............................................... 120
    sedge grass .........................................63
    switchgrass................................120, 172
green ash tree .........................................185
greens................................................... 207
    collards ............................................. 13
    kale ..................................13, 15, 146
    spinach............................................. 146
*Grow Organic* ........................................... 87

# H

hackberry ...............................................185
hardy geranium ....................................... 111
hardy hibiscus .........................................89
hawthorne.............................................. 207
hemerocallis ............................................83
hemlock........................................ 176-177, 180
    Eastern hemlock .................................185
herbicides......32, 60, 66, 153, 154, 171, 209, 213
    pre-emergent herbicide............ 83, 209-210
    select kill herbicide ............................209
    total kill herbicide ........................83, 209
heuchera..................................................24
hibiscus (see also rose mallow)
    hardy hibiscus .....................................89
hickory..................................................206
    shagbark hickory ............................... 182
hollies ............................................140, 172
    American Holly................................. 140
    deciduous holly ................................ 140
    Yaupon holly.....................................185
hollyhock................................................ 52, 60
    Nigra hollyhock ................................. 190
honeydew melons...................................137
honeysuckle
    bush honeysuckle ..............................213
    Japanese honeysuckle .......................... 212

trumpet honeysuckle ................47, 186, 212
hornworm ............................. 87, 100, 101, 123
hostas...... 38, 49, 52, 56, 62, 63, 80, 81, 83, 115, 180, 195
Hot Pepper Wax .........................................39
hubbard squash ........................................98
hyacinth bean ................................. 126, 190
hyacinth ................................57, 142, 145, 149
    grape hyacinth .................... 23, 57, 142, 190
hybrid tea rose ........................ 16, 33, 51, 81
hydrangea...... 63, 69, 80, 117, 160, 166-169, 190, 197, 204, 205
    climbing hydrangea .......................124, 126
    hydrangea Annabelle ............................. 111
    hydrangea arborescens......................... 166
    hydrangea macrophylla ....................... 166
    hydrangea paniculata ........................... 166
    hydrangea quercifolia ................... 166, 205
    hydrangea tardiva ................................. 180
    lacecap hydrangea............................37-38
    oakleaf hydrangea ................ 23, 24, 48, 147
hyssop ..................................................... 190

# I

ilex ............................................................63
imidacloprid........................................102, 177
impatiens ................... 48, 50, 63, 81, 180, 207
Independent Garden Center Association ...27
Indigo rose ............................................... 190
inkberry .................................................. 120
insecticides ........ 15, 33, 60, 85, 86, 93, 99, 116, 117, 177
insects ... 39, 45, 60, 86-87, 130-131, 176, 177, 210
*Insects and Gardens: In Pursuit of a Garden Ecology* ................................................................ 130
invasive plants....................................211-213
irises ...........24, 52, 53-55, 56, 83, 120, 180, 195
    blue flag iris................................ 54, 55
    Dutch iris ........................................ 23, 145
    German bearded iris ....... 23, 26, 53, 54, 55, 56, 190
    Japanese iris .................................54, 120
    Siberian iris ....................23, 54, 120, 190
    yellow flag iris .......................... 54, 55, 120
itea .............................................................63
    dwarf itea .................................................91
ivy .......................................................62, 124
    English ivy.....................................186, 212

# J

Jane magnolia ........................................56, 83
Japanese andromeda................................ 180
Japanese barberry .................................... 212
Japanese beetles ...................................... 102
Japanese cedar......................................... 180
Japanese fern ............................................80
Japanese honeysuckle .............................. 212
Japanese iris .......................................54, 120
Japanese maple............................83, 180, 185
Japanese stewartia ....................................183
Jekyll, Gertrude ................................. 24, 56
joe pye weed .......................................... 120
Judas tree (see also redbud) ......................43
June bearers ....................................... 64, 65
juniper .............................................180, 185
Justice, Monty .................................... 34-36

# K

kale ...............................................13, 15, 146
katsura tree ..............................................183
Kentucky coffee tree ..............47, 162, 181, 206
Kentucky Exotic Plant Council .......... 212-213
Kingsolver, Barbara ....................................30
knautia .................................................... 111
knockout roses .............................. 33, 38, 48
kohlrabi ............................................... 13, 15
Korean lilacs ............................................. 41
kousa dogwoods..................................82, 140
kudzu ........................................................211
Ky. 31 fescue ........................................... 212

# L

lacebark elm ....................................182, 183
lacebark pine ............................................183
lacecap hydrangea .............................37-38
lambs quarter ......................................... 212
lantana .................................................... 117
Large Red cherry tomato ......................... 190
lavender ............................................ 24, 26
    English lavender ................................. 190
lawns ................................... 32, 153-154
lespedeza ................................................ 212
lettuce ................................................92, 146
    Amish Deertongue lettuce.................... 190
leucothea ..................................................63
Leyland cypress ........................................174
licorice vine.............................................. 180
lilacs............................... 24, 37, 40-42, 147, 190
    Korean lilacs ...................................... 41
lilies (see also daylily) ........................132-134
    Asiatic lily ........................................ 160
    blackberry lily .........................................48
    Madonna lily ...............................132-133
    summer lily........................... 89, 132-134
Listada De Gandia eggplant ..................... 190
loam ............................................. 28, 208
lobelia ......................................................24
locust tree ................................................50
London plane tree ....................................183
love-lies-bleeding ..................................... 190

# M

Madonna lily ......................................132-133
magnolia................... 57, 82, 157-158, 180, 205

cucumber magnolia..................157, 163, 183
Jane magnolia ............................56, 83, 158
magnolia acuminata..............................158
magnolia denudata................................158
magnolia soulangeana...................158, 205
magnolia stellata.....................................158
saucer magnolia......................................158
Southern magnolia....................139, 157-158
star magnolia..........................................158
sweet bay magnolia ........................120, 158
mahonia .........................................................63
maidenhair grass ........................................ 180
maidenhair tree ............................................46
malathion.............................................101, 102
mallow (see also hollyhock).................52, 105
malva............................................................105
maples ................................................. 62, 206
amur maple ...........................................183
Japanese maple..........................83, 180, 185,
Norway maple.........................................62
paperbark maple................................ 182
red maple .....................................120, 162
sugar maple............................................62
marigolds ......... 48, 50, 57, 60, 67, 108, 117, 118, 186, 207
French marigold .................................. 180
marsh marigold ................................... 120
yellow marigold......................................92
meadowsweet ............................................ 120
melons.................................. 58, 59, 135-137
cantaloupe............................... 101, 135-137
honeydew melon .................................137
muskmelons ................................. 101, 137
watermelon ...........................39, 135-137
Mexican sunflower .......................................48
millet...............................................................24
mimosa........................................................ 212
mint ............................................................. 103
miscanthus....................................................172
mission bells ............................................... 180
monkshood ............................... 63, 67, 69, 80
Monty's Joy Juice................................... 34-36
moon garden ............................................. 216
moon signs........................................... 214-215
moonflower........................................190, 216
morning glory....................................... 47, 126
mountain ash tree......................................185
mulch.............................................154, 156, 159
mulberry
white mulberry..................................... 212
multiflora rose .................................... 211, 212
mums .......................................... 24, 56, 57
muskmelons ....................................... 101, 137

# N

nandina ......................................................56
narcissus ................................................ 150

nasturtium ................................60, 180, 190
Neem............................................................86
nepeta (see also catnip, catmint) ............ 207
nicotiana (see also flowering tobacco). 48, 180
Nigra hollyhock ......................................... 190
Norway maple ...........................................62

# O

oak trees....................................................204-206
bur oak .......................................... 162-163
pin oak ...................................................185
red oak ........................................ 181, 185
white oak....................................... 182, 204
willow oak............................................. 180
oakleaf hydrangea.................... 23, 24, 48, 147
okra........................................................94-96
orchid................................................. 198-199
organic gardening................................. 85-87
ornamental grass ................ 147, 172, 201, 207
Oxford Blue love-in-a-nigh ..................... 190

# P

pachysandra.................................................62
Paeonia lactiflora .....................................155
painted fern..................................................63
painted tongue ........................................ 190
panicum ............................................110, 111
paperbark maple................................. 182
Peace rose.....................................................35
peaches ............................................. 74, 195
pears ..........................................................74
peonies.......... 49, 52, 53, 147, 155-156, 196, 207
Festiva Maxima peony ......................... 190
peppers ................................................. 38-39
petunia................23, 24, 50, 69, 88, 92, 93, 108
phlox..................................................24, 89, 205
summer phlox................................ 180, 205
pieris..............................................................63
pigweed .......................................................59
pin oak ........................................................185
pine trees ................................................. 140
lacebark pine.........................................183
umbrella pine ..................................... 180
white pine ............................................172
pinks ..........................................................185
pitcher plant............................................. 120
plums ..........................................74, 185, 195
plumbago ................................................. 180
popaver .......................................................83
poppy (see California poppy)
Portland rose ............................................. 190
possum haw ............................................. 120
potentilla....................................................118
pre-emergent herbicide ..................... 209-210
pruning .......................................................76
pumpkins ....................................................98
purple basil .................................................92

purple loosestrife .......................... 211, 212, 213
pyracantha (see also firethorn) .......... 172, 186
pyrethrins ................................................... 86

## Q

Queen Anne's lace ...................................... 86

## R

raspberries ............................................... 148
Red Fig tomato ......................................... 190
red maple ......................................... 120, 162
red oak ............................................. 181, 185
red-osier dogwood ............................ 120, 185
red spider zinnia ...................................... 190
redbud ................................. 43-45, 47, 207
    Eastern redbud ...................................... 180
redwood
    dawn redwood .................................... 163
*Restoring American Gardens* ...................... 188
rhizomes ............................................... 54-55
rhododendron ...................... 20-22, 42, 194
ribbon grass ............................................... 69
river birch ......................................... 50, 120
    dwarf river birch ................................... 83
roses ........... 16-19, 34-36, 42, 49, 56, 57, 68, 89, 147, 174
    Bella Donna rose ................................. 190
    Charles De Mills rose .......................... 190
    climbing rose ............................... 124, 125
    hybrid tea rose ...................... 16, 33, 51, 81
    Indigo rose .......................................... 190
    knockout roses ....................... 33, 38, 48
    multiflora rose .............................. 211, 212
    Peace rose ............................................ 35
    Portland rose ....................................... 190
    shrub roses .......................................... 207
rose mallow (see also hibiscus) ................ 120
rose of Sharon ............................................ 47
rotenone ............................................. 86, 99
rudbeckia (see also black-eyed Susans) ..... 48, 69, 89, 117, 118, 147, 207
rushes ...................................................... 120
Russian sage ........................................ 24, 48

## S

sage ................................... 87, 93, 106-108
    annual sage .......................................... 24
    clary sage (see also Salvia sclarea) ........ 107
    Russian sage .................................. 24, 48
    scarlet sage ......................................... 107
salmonella .................................................. 27
salvia ..... 23, 24, 26, 50, 52, 60, 68, 83, 106-108, 131, 180, 197, 207
    Salvia divinorum .......................... 106, 108
    Salvia farinacea .................................. 107
    Salvia guaranitica ................................ 111
    Salvia nemerosa .................................. 107

Salvia pratensis ........................................ 108
Salvia sclarea (see also clary sage) ........ 107
Salvia splendens .................................... 107
Salvia superba ....................................... 107
sand ......................................................... 208
    builder's sand ................................. 18, 21
saucer magnolia ....................................... 158
scabiosa ................................................... 185
scallop squash ........................................... 98
scarlet sage .............................................. 107
sedge grass ................................................ 63
sedum ................................... 68, 180, 207
select kill herbicide .................................. 209
semi-dwarf fruit trees ................................. 75
serviceberry ........................ 83, 185, 186, 207
Sevin dust ............................... 99, 102, 123
shade plants ......................................... 61-63
shagbark hickory ..................................... 182
*Shaker Medicinal Herbs* ............................. 188
Shasta daisy ...................................... 69, 152
Shooting Star Nursery .............................. 33
shrub roses ............................................. 207
shrubs . 32, 33, 37, 47, 62, 63, 81, 82, 83, 117, 147, 160, 185, 207
Siberian iris ....................... 23, 54, 120, 190
silver lace vine .......................................... 47
snapdragon ............................................... 38
soil ............................................. 151-152, 208
soil pH .................................................... 208
soil test ........................................... 151, 152
sourwood ................................................ 183
Southern magnolia ................... 139, 157-158
spaghetti squash ....................................... 98
Spear and Jackson ................................... 201
species tulips .......................................... 190
spicebush ......................................... 120, 185
spiderwort ................................................ 52
spinach ................................................... 146
squash ............................... 97-99, 101, 207
    acorn squash ....................................... 98
    banana squash .................................... 98
    butternut squash ................................. 98
    hubbard squash .................................. 98
    scallop squash ..................................... 98
    spaghetti squash ................................. 98
    summer squash ................................... 97
    winter squash ..................................... 98
    zucchini ............................. 50, 60, 97-98
St. John's wort ......................................... 212
stacys ........................................................ 83
star magnolia .......................................... 158
statice ..................................................... 185
stinging nettle ........................................... 62
stinkbug ................................................... 99
stonecrop ................................................. 48
strawberries .............................. 64-66, 148
    day-neutral strawberries ..................... 65

ever-bearing strawberries ......................65
strawflower ............................................. 190
stumpery..................................................80
sugar maple .............................................62
sulfur........................................................22
sumac..................................... 47, 139, 140, 185
summer lilies................................ 89, 132-134
summer phlox ................................ 180, 205
summer squash.........................................97
sunflowers....................... 60, 88-90, 147, 186
    Mexican sunflower ................................48
sutera ..................................................... 180
sweet autumn clematis ........................47, 112
sweet azalea .............................................22
sweet bay magnolia ..........................120, 158
sweet corn........................ 70, 72, 73, 100, 135
sweet peas............................................... 190
sweetgum tree ................................. 185, 206
switchgrass .................................... 120, 172
sycamore ............................................... 182
Syringa vulgaris ....................................... 40

# T

tansy ........................................................86
taxus (see also yew) ........................... 172, 174
thalia daffodil...........................................57
*The Organic Gardener's Handbook of Natural Insect
and Disease Control* ............................................ 87
*The Right Rose in the Right Place* .....................188
*The Southern Living Garden Book* ....................188
thyme............................................ 87, 93, 190
tickseed (see also coreopsis)...................... 51
tomatoes .... 39, 50, 58, 59, 60, 92, 101, 121-123,
    127-129, 146, 207
    Brandywine tomato ...........39, 123, 127, 128,
    129, 190
    cherry tomato ..............................92, 93, 128
    Large Red cherry tomato ...................... 190
    Red Fig tomato ..................................... 190
tomato hornworm ....................................101
tools ................................................ 200-201
total kill herbicide ..............................83, 209
tree of heaven ........................................ 212
trillium.....................................................80
true geranium.........................................207
trumpet creeper........................................47
trumpet honeysuckle ....................47, 186, 212
trumpet vine........................................... 126
tulips.............. 23, 57, 62, 67, 142, 149, 150, 190
    species tulips........................................ 190
tulip poplar ............................47, 161, 163, 185

# U

umbrella pine ......................................... 180

# V

vegetable garden......................... 27-29, 58-60

verbena ................................................... 180
viburnum .. 22, 23, 24, 47, 63, 117, 140, 147, 172,
    185, 186, 190, 204, 207
vinca ................................................50, 180
vines ............................ 47, 112-113, 124-126
    crossvine ................................................47
    licorice vine ......................................... 180
    silver lace vine .......................................47
    trumpet vine ........................................ 126
viola ........................................................ 38
virgin's bower...........................................112
Virginia bluebells ......................................62
Virginia creeper ..................................... 186
Virginia sweetspire................................. 120
vitex ...................................................... 147

# W

water tupelo ........................................... 120
watermelon ................................... 39, 135-137
    Dixie Queen watermelon ....................... 39
wax privet................................................117
white birch...............................................181
white mulberry...................................... 212
white oak ....................................... 182, 204
white pine ..............................................172
Whitehall.................................................78
Wiche, Fred ........................................... 109
Wiche, Jeneen ........................... 108, 109-111
willow................................................50, 117
willow oak ............................................. 180
winter squash...........................................98
wintercreeper ................................... 126, 211
wisteria ..............................................24, 104
    climbing wisteria ................................ 125
witchhazel...............................................185
wooly adelgid .................................. 176-177
worms ....................................................207

# Y

yarrow (see also achillea) ......... 52, 67, 69, 86,
    87, 117
Yaupon holly............................................185
yellow flag iris ................................ 54, 55, 120
yellow marigold.........................................92
yew (see also taxus) ..................................56
yucca ......................................................83

# Z

zinnias............38, 48, 57, 60, 108, 117, 118, 147,
    186, 207
    red spider zinnia ................................. 190
zucchini ........................................50, 60, 97